ENVIRONMENTAL JUSTICE

Selected Titles in ABC-CLIO's
CONTEMPORARY
WORLD ISSUES
Series

For a complete list of titles in this series, please visit
www.abc-clio.com.

Books in the Contemporary World Issues series address vital issues in today's society, such as genetic engineering, pollution, and biodiversity. Written by professional writers, scholars, and nonacademic experts, these books are authoritative, clearly written, up-to-date, and objective. They provide a good starting point for research by high school and college students, scholars, and general readers as well as by legislators, businesspeople, activists, and others.

Each book, carefully organized and easy to use, contains an overview of the subject, a detailed chronology, biographical sketches, facts and data and/or documents and other primary-source material, a directory of organizations and agencies, annotated lists of print and nonprint resources, and an index.

Readers of books in the Contemporary World Issues series will find the information they need to have a better understanding of the social, political, environmental, and economic issues facing the world today.

ENVIRONMENTAL JUSTICE

A Reference Handbook

Second Edition

David E. Newton

**CONTEMPORARY
WORLD ISSUES**

A B C 🌐 C L I O

Santa Barbara, California
Denver, Colorado
Oxford, England

Library of Congress Cataloging-in-Publication Data

Newton, David E.
Environmental justice : a reference handbook / David E. Newton. — 2nd ed.
    p. cm. — (Contemporary world issues)
  Includes bibliographical references and index.
  ISBN 978-1-59884-223-4 (hard copy : alk. paper) —
ISBN 978-1-59884-224-1 (ebook)   1. Distributive justice.
2. Environmental justice—Social aspects.   3. Civil rights.   I. Title.

JC578.N485   2009
363.7—dc22

                                    2008044400

13 12 11 10 09   1 2 3 4 5

ABC-CLIO, LLC
130 Cremona Drive, P.O. Box 1911
Santa Barbara, California 93116–1911

This book is also available on the World Wide Web as an ebook.
Visit www.abc-clio.com for details.

This book is printed on acid-free paper ∞

Manufactured in the United States of America.

Once again,
for Delores and Reg
with as much love as ever!

Contents

List of Tables

Preface

Environmentalism and the civil rights movement have been two of the most important social movements in the United States over the past half century. At first glance, these two great movements would appear to have little or nothing in common. And, in fact, some observers have argued that steps taken to improve the nation's environmental quality may actually have had a deleterious effect on the social and economic status of African Americans, Hispanic Americans, Asian Americans, and Native Americans. Studies have shown, for example, that laws passed to reduce environmental impacts overall have, in some cases, had an unfair impact on minority or poor communities.

Only within the last three decades has it become apparent that environmentalism and civil rights do have a great deal in common. That realization has come about primarily as the result of redefining the "environmental issues" to which Americans should turn their attention. More and more individuals and groups are pointing to the special environmental problems facing minority and low-income communities: hazardous waste sites, polluting industries, occupational hazards, and exposure to lead and other toxic metals, for example. The emphasis seems to be shifting from an almost exclusive concern with issues such as wilderness preservation and protection of endangered species to maintenance of livable environments for people of color and low-income communities.

Out of this view of environmental issues has grown a new movement—the environmental justice movement—that attempts to analyze patterns of disproportionate exposure to environmental hazards experienced by minority and low-income communities, to understand how such patterns have developed, and to develop programs by which disproportionate exposures can be remedied and prevented.

The environmental justice movement has undergone a significant evolution in the last three decades. At first, the concerns expressed by its leaders were largely ignored or regarded with little concern. Low-income and minority communities themselves seldom had the knowledge or the skills to deal with the special environmental inequities with which they were confronted. Over time, however, that situation changed. Strong leaders from the African American, Asian American, Native American, Hispanic American, and other communities began to develop theory and practices on which to base a strong and effective environmental justice movement. In some ways, a turning point for the movement was reached in 1994, when President Bill Clinton issued an executive order requiring all federal agencies to make environmental justice an integral part of their missions.

Progress in dealing with environmental inequities has not, however, been smooth or constant. It has fallen prey to the influence of varying social, economic, and political forces that have (and often have not) supported the growth of efforts to achieve environmental justice. During the administration of President George W. Bush, for example, much of the momentum created by President Clinton's executive order was lost, as Bush officials stepped back from vigorous enforcement of environmental policies and regulations and, indeed, attempted to redefine the meaning of environmental justice itself.

This book examines the nature, growth, and vicissitudes of the environmental justice movement. Chapter 1 provides a general introduction to the history of the movement, explaining the forces that led to its birth, the philosophy that underlies much of its work today, and the tools through which it attempts to achieve its goals. The chapter also contains a number of examples of cases in which environmental inequities have affected individual communities.

The irony of the modern environmental justice movement is that it carries such great hope for low-income people and minority communities in the relief of the environmental hazards to which they are often exposed. But concrete success in dealing with many of the issues has often seemed too slow and much too inadequate. Chapter 2 reviews some of the problems and controversies that may have contributed to this situation. It also discusses some possible solutions to the problems that hinder the development of the movement.

One might easily think of environmental justice as a uniquely American movement grounded in environmental problems found exclusively in the United States. And, in some ways, that view is accurate. Many of the problems faced by low-income and minority communities (and by Americans as a whole) result from the prodigious industrial complex operating in this country, with its consequent effluents and wastes constantly being released into the environment. But people in other parts of the world face similar problems of inequitable exposure to hazardous materials based on income and race. For example, industries in developed nations have long viewed less-developed nations as legitimate dumping grounds for the wastes they produce at home. Chapter 3 considers some issues related to environmental inequities in parts of the world other than the United States.

Chapters 4 through 8 provide a variety of resources to help the reader continue further study of environmental justice. Chapter 4 contains a chronology of the movement, a movement that, although still young, has made some important strides. Chapter 5 consists of biographical sketches of some important figures in the movement. Chapter 6 contains documents relating to the growth and work of the environmental justice movement, including relevant laws, recommendations for action, and proposed legislation.

Chapter 7 provides a list of organizations active in the field. The list is representative only; one characteristic of the environmental justice movement is its emphasis on action by local and regional groups, which precludes the ability to provide an exhaustive list. Chapter 8 contains a selected bibliography of books, articles, reports, and Internet sources relating to the environmental justice movement. A glossary of terms commonly used in dealing with environmental justice issues is included, and an index is also provided.

1

Background and History

Warren County, North Carolina: September 15, 1982

The protest tactics seemed clear enough. Most residents of Warren County knew how civil rights battles of the 1960s had been fought: with civil disobedience and nonviolent protest. The issue in 1982 was somewhat different, involving the construction of a waste disposal dump for toxic chemicals, but poor African Americans in Warren County knew their history well enough. The road to follow was not one of violent confrontation but of nonviolent demonstration and protest.

The issue had arisen four years earlier when the Ward Transformer Company of Raleigh had illegally and surreptitiously disposed of 31,000 gallons of toxic polychlorinated biphenyls (PCBs) along 240 miles of roadways in 14 North Carolina counties. When the state of North Carolina uncovered the action, it was faced with the problem of digging up and relocating 40,000 cubic yards of contaminated soil. The state decided to bury those wastes in rural Shocco Township in Warren County.

The state's decision came under criticism almost immediately. Geological studies showed that soil under the proposed landfill was not impervious to leaching from the landfill, as required by U.S. Environmental Protection Agency (EPA) regulations, and the water table in the area was only about 7 feet below the landfill bottom, 43 feet shallower than required by the EPA. Since the majority of Warren County residents obtain their water from wells, these geological conditions raised serious questions

1

about the state's decision. On June 4, 1979, however, the EPA gave North Carolina a waiver on both of these requirements and issued a permit for the landfill.

Residents of Warren County had had little experience in fighting political battles. As a group, they were poor (the county ranked 97th among North Carolina's 100 counties in per capita income) and African American (75 percent in Shocco Township and 64 percent in Warren County) and lacked the resources, knowledge, and skills to fight a state political and bureaucratic system. Still, they used the skills they had. They made public appeals, called for help from national leaders of civil rights and environmental groups, and, in the end, prostrated their bodies in front of trucks carrying PCB-contaminated soil to their community. In all, 523 protestors were arrested.

In one sense, the Warren County protest failed. Over a six-week period, the 20-acre landfill was filled with 7,223 truckloads of wastes. The wastes remained in place for nearly 20 years, all the while leaching contaminants into the underlying soil. In a larger sense, however, the protest was more successful than anyone could have imagined. For the first time in history, poor African Americans banded together—with the support of civil rights and environmental groups on a national level—to fight an environmental problem affecting a poor minority community.

Environmental Inequities, Environmental Racism, and Environmental Justice

Most great social movements in history can be marked by a "defining moment," a single event so powerful and moving that later generations can look back and say, "It all started at . . . ." Rosa Parks's refusal to give up her seat on a Montgomery, Alabama, bus to a white man and move to the back of the bus is one such example of a defining moment in the civil rights movement. Historians may well look back at the protest by Warren County citizens and say, "That protest was the defining moment of the environmental justice movement."

Of course, such views of history are usually overly simplified and often incorrect. A nascent civil rights movement had been active long before Parks's heroic decision to confront racist laws in Montgomery. And the seeds of a movement that would confront environmental inequities had been planted years before the Warren County protest brought the movement to public attention.

The issue facing residents of Warren County, North Carolina, in 1982 was and is not unique in the United States. The issue has been described as *environmental inequity or environmental racism*, terms that refer to the generally accepted evidence that environmental hazards are not distributed equally among various groups of people, either in the United States or throughout the world. Instead, communities of color and, to a lesser extent, poor people in general are exposed to hazardous and toxic wastes, dangerous working conditions, polluted air and water, and other environmental insults to a greater degree than are noncolored communities and people of higher economic status. Although the reasons for this pattern are disputed, its existence has been confirmed by multiple research studies. The term *communities of color*, as described by those active in the environmental justice movement, refers to people who are nonwhite: African Americans, Hispanic Americans, Asian Americans, and Native Americans, for example.

The terms used to talk about this phenomenon are important because they focus on and emphasize quite different aspects of the issue. *Environmental inequity* refers to a geographic reality, a pattern in which hazardous waste sites, polluting industries, nuclear waste dumps, and other environmental threats are more likely to be located within or adjacent to communities of color or poor communities. The existence of environmental inequities in the United States, within individual states, or in other localities can be determined—at least in principle—by research studies. Researchers count the number of environmentally hazardous facilities and determine whether inequities are likely to occur in such communities. Such studies have been conducted, and the evidence implies that environmental inequities do exist in the United States. For example, researchers at the Southeast Community Research Center in Atlanta reported in 2007 on the use of agricultural pesticides in and near communities of color in the rural U.S. South. They found that such communities were exposed to greatly disproportionate numbers of pesticide use events compared to comparable communities of white residents (Griffith, Tajik, and Wing 2007). A number of similar studies, along with some that dispute the existence of environmental inequities, are discussed later in this chapter.

Environmental racism is a term that introduces a second concept into the discussion of environmental inequities. Environmental racism goes beyond acknowledging that such inequities exist and suggests a reason for their existence: racism. The phrase environmental racism was first used in 1982 by Benjamin Chavis, then executive director of the National Association for the Advancement of Colored People (NAACP), in connection with the protest at Warren County. He later explained precisely what he meant by the term during testimony before the U.S. House of Representatives Subcommittee on Civil and Constitutional Rights on March 3, 1993: "Environmental racism is defined as racial discrimination in environmental policy making and the unequal enforcement of environmental laws and regulations. It is the deliberate targeting of people of color communities for toxic waste facilities and the official sanctioning of a life-threatening presence of poisons and pollutants in people of color communities. It is also manifested in the history of excluding people of color from the leadership of the environmental movement" (U.S. Congress 1993, 6).

A second term, *environmental discrimination*, is also used in referring to the unequal distribution of environmental insults but may suggest that such inequalities affect communities other than those of color, such as those of low income. An important feature of both environmental racism and environmental discrimination is that they suggest that environmental inequities occur not because of chance, random events in history but as the concrete and specific consequences of official public and corporate policies, whether conscious and deliberate or not.

Proving the existence of environmental racism or environmental discrimination is a quite different matter from proving the existence of environmental inequities. Even if one grants that such inequities exist, it does not necessarily follow that they are the result of some specific intent on the part of an individual or a corporation. The debate as to whether environmental inequities occur as the result of concrete policies and practices designed to bring them about or as the incidental consequences of other business practices will be reviewed later in this book.

The term *environmental justice* alludes to yet another aspect of this issue. It refers to policies and practices by which existing environmental inequities can be corrected and prevented in the future. It focuses on research programs that attempt to detect the existence of environmental racism and environmental discrimination; that uncover the underlying reasons that hold such practices in place; and that promote the enforcement of existing laws and regulations, the adoption of new rules and regulations, and the changes in philosophies and attitudes needed to eliminate environmental racism and environmental inequities from society. The achievement of environmental justice among communities is also called *environmental equity,* the situation in which environmental risks and benefits among all groups of people are approximately comparable.

Dr. Robert D. Bullard, one of the most prolific and articulate writers on the subject of environmental justice, has identified three broad categories into which the field of environmental justice can be subdivided. These are procedural equity, geographic equity, and social equity. Procedural equity refers to questions of fairness, "the extent to which governing rules and regulations, evaluation criteria, and enforcement are applied in a nondiscriminatory manner." Geographic equity refers to the location of environmental hazards with regard to communities of color and poor communities. Social equity concerns the way in which social factors, such as race, ethnicity, class, and political power, have an impact on and are reflected in environmental decision making (Bullard 2000, 116).

The environmental inequities that exist in the United States include a variety of hazardous conditions. During the 1970s, much attention was focused on air and water pollution. Data from that period clearly show that the impact and costs of these environmental problems—although often not explicitly stated in reports—were not shared equally by all Americans. Two decades later, in the early stages of the environmental justice movement, research focused on municipal landfills, hazardous waste dumps, and emissions from chemical plants. As the movement has matured, however, it has become clear that many other forms of environmental inequities exist: in the siting of nuclear waste sites; in the nature and extent of occupational hazards to which workers are exposed; in the health hazards to which people are exposed in their daily diets; in planning for and construction of highways and public transit systems; in the siting of parks and other recreational facilities; in the production, testing, and disposal of conventional weapons systems; in the international distribution of hazardous and toxic wastes; and in the manufacture and sale of hazardous products.

Examples of Environmental Inequities

The term *racism* may cause one to focus on the inequalities that African Americans have historically experienced in the United States. It is clear, however, that other groups within American society—including Native Americans, Asian Americans, Pacific Islanders, and Hispanic Americans—have also been subjected to similar kinds of discrimination. The examples on the following pages are a few of the hundreds that could demonstrate the environmental insults to which minority and low-income groups within American society have been exposed. The cases illustrate the disproportionate exposure of minorities to environmental hazards; however, they do not in and of themselves necessarily prove the existence of environmental racism.

West Dallas, Texas

In the late 1980s, West Dallas was a predominantly poor African American section of Dallas, Texas, with a population of 13,161, of whom 85 percent were African American. In recent decades, the area has grown considerably (2000 population: 24,132) and become primarily Hispanic American. But it remains one of the poorest regions of Dallas. When the area was annexed by Dallas in 1954, a number of private homes were torn down as part of a "slum clearance" program to make way for large public housing projects. As a result of this program, the Dallas Housing Authority became the largest single landlord in West Dallas.

Immediately adjacent to the housing authority's 3,500-unit West Dallas housing projects is the 63-acre West Dallas RSR lead smelter, originally the Murph Metals secondary lead smelter. At its peak operation in the 1970s, the smelter released more than 269 tons of lead into the air annually. A significant fraction of the lead was blown across the homes and recreational areas of West Dallas.

In theory, such emissions came under the control of a strong lead emissions ordinance passed by the Dallas city council in 1968. According to Robert D. Bullard, however, "city officials systematically refused to enforce [the city's] lead emission standards" (Bullard 2000, 47). In support of this observation, studies found blood lead levels in West Dallas children to be 36 percent higher than those in children in control areas, a result attributed to the children's exposure to smelter emissions.

The period between the mid-1960s and the early 1980s was marked by the continuing release of lead particulates from the smelter, multiple studies showing elevated blood lead levels in West Dallas children. Ongoing efforts by citizens of West Dallas to gain relief from their exposure to polluted air met with persistent stonewalling by local and national officials in the enforcement of existing regulations.

The issue was resolved in 1983, when the state of Texas and the city of Dallas sued RSR for violations of city, state, and federal lead emission standards. The case was settled out of court when RSR agreed to clean up soil in the West Dallas area that had been contaminated by lead emission from its smelter, to conduct a blood lead screening program for children and pregnant women in West Dallas, and to install antipollution equipment on its West Dallas plant. The final point of the agreement became moot four months later when the smelter was closed permanently.

The final word in this story may have been written in 1985, when RSR also agreed to a settlement in a civil case brought on behalf of 370 children in West Dallas. The company agreed to a program of payments that eventually reached nearly $45 million paid to those harmed by its smelter's emissions. Although the cash settlement was large, it can never, as Bullard has pointed out, "repay the harm caused by lead poisoning of several generations of West Dallas children." On the other hand, it does send a message, he says, "that poor black communities are no longer willing to accept other people's pollution" (Bullard 2000, 51).

Churchrock, New Mexico

Churchrock Chapter is a governmental unit within the Navajo Nation, the largest Native American tribe in the United States. The chapter is located east of Gallup, New Mexico, in an area characterized by extremely dry conditions with an average annual rainfall of about seven inches. The major source of water in the region is the Rio Puerco, a stream that runs only intermittently, when fed by snow runoff or rainstorms.

Churchrock is the site of the longest continuous period of uranium mining in the Navajo Nation. That period began in 1954 with the discovery of uranium deposits along the north edge of the Rio Puerco Valley and continued until 1986. Mining rights were leased to uranium mining companies by the Navajo tribal government, but without the consent or participation of individual Navajo families living in the area. During most of the 30-year period during which mining continued, residents of the Churchrock Chapter had relatively little information about the environmental effects caused by uranium mining and milling in their area.

One of the most important of those effects was the removal of water from the Morrison Formation aquifer near the head of the Rio Puerco. Mining companies pumped water out of the aquifer at the rate of 5,000 gallons per minute to support the construction and operation of underground mines for the removal of uranium. That water was then returned to the Rio Puerco, converting the river into a continuously flowing stream.

This practice had two important effects on the Churchrock environment. First, it resulted in a significant loss of underground water from which many people drew water by means of wells. Second, it resulted in the release of water contaminated with radioactive wastes into the Rio Puerco, the region's major water source.

For years, the two companies mining uranium in the Churchrock region—the Kerr-McGee Nuclear Corporation and United Nuclear Corporation—argued that the Federal Water Pollution Control Act did not apply to their operations since the Rio Puerco was located on Native American lands and was not, therefore, a part of the United States. It was not until 1980 that the courts confirmed that the two companies were subject to the same Clean Water Act jurisdiction as were companies operating in other parts of the United States.

In 1983, the staff of the Southwest Research and Information Center in Albuquerque examined Kerr-McGee and United Nuclear records to determine the extent to which the two companies were in compliance with federal Clean Water Act regulations. They found that Kerr-McGee was out of compliance 7 of the 35 months studied, and United Nuclear was out of compliance 13 of 38 months in one location and 25 of 37 months in a second location. "Out of compliance" meant that mines were releasing waters that contained a higher concentration than permitted of dissolved uranium and/or radium-226, carried an excess of suspended solids, had a pH (a measure of the water's acidity or alkalinity) that was higher or lower than permitted, or failed to meet other water quality standards.

As unsatisfactory as these records were, they were dwarfed by a serious accident: the failure of a United Nuclear uranium mill tailings dam on July 16, 1979. More than 94 million gallons of contaminated liquids with a pH of 1 (roughly comparable to that of battery acid) were released into the Rio Puerco. The spill posed a serious health threat not only to Native Americans living in the area, but also to the livestock upon which many Native Americans depended for their livelihood. In 1985, United Nuclear settled a lawsuit brought as a result of this accident by agreeing to pay $550,000 to 240 plaintiffs, an award of about $2,000 per plaintiff.

Still, it was not the 1979 spill that has been the most serious problem for residents of the Churchrock area. William Paul Robinson, research director of the Southwest Research and Information Center, has written that the spill "does not appear to have had as devastating an effect on the Rio Puerco as the decades of mine dewatering (the removal of water that seeps into a mine as it is dug), which preceded the spill. Studies of human and livestock health effects after the spill indicated that the same pollutants found in high concentrations in mine and tailings water had shown up in abnormally high levels in the muscles and organs of cattle, sheep, and goats that grazed along the Rio Puerco downstream of the mines and mills" (Robinson 1992, 160).

As is often the case in environmental justice cases, actions against uranium mining in the Churchrock area may have been more successful than one might have imagined. Residents have formed the Puerco Valley Navajo Clean Water Association to learn more about the dangers of polluted waters and to search for new and safer water supplies for their community. As Robinson has pointed out, the Association's 10th anniversary celebration on July 16, 1989, "focused on the need for water in the communities, not the horror of the spill itself" (Robinson 1992, 161).

"Cancer Alley"

In some instances, a pattern of environmental inequity may extend over broad regions, affecting many communities at once. Such is the case with an 85-mile stretch of land along the Mississippi River between Baton Rouge and New Orleans, Louisiana. Louisiana has long publicized itself as a "sportsman's paradise" and depended on agriculture and fishing as its main sources of income. At the end of World War II, however, the state's political economy began to change. Attracted by the potential for easy transportation and waste disposal as well as the availability of cheap labor, chemical and petroleum companies moved into the region between the state capital and its largest city. Over the decades, these companies dramatically changed the character of the region.

Today, more than 125 companies producing fertilizers, gasoline, paints, plastics, and other chemical products line the banks of the Mississippi. Approximately one-quarter of the nation's petrochemicals are produced in the region. In addition, the plants release huge amounts of toxic wastes into the air, water, and surrounding countryside. By one estimate, more than 2 billion pounds of toxic chemicals were emitted in the two-year period between 1987 and 1989 (Louisiana Advisory Committee 1993). For this reason, the region has been given the name "Cancer Alley" by residents of the area. The *Washington Post* has referred to the region as a "massive human experiment" and "a national sacrifice zone" (Maraniss and Weisskopf 1989, A1).

The development of the new industrial complex along Cancer Alley, while undoubtedly beneficial to the state economy, has caused severe dislocations among communities along the river. In a 1993 study of Cancer Alley, the Louisiana Advisory Committee to the U.S. Commission on Civil Rights selected eight communities to study in detail: Revilletown, Sunrise, Morrisonville, Alsen, Wallace, Forest Grove, Center Springs, and Willow Springs. The committee found that the interaction between corporations and communities differed from town to town, and the ultimate resolution of disputes between the two entities also differed from case to case.

In some instances, residents were able to prevent a new plant from moving into an area. In such cases, the community was able to survive, although it was not necessarily spared a significant increase in exposure to air- and water-borne pollutants released by nearby plants. In other cases, towns acceded to offers from companies to sell their homes and land to make room for new plants. Two of the communities studied, Revilletown and Sunrise, were dismantled completely as a result of buyout programs, and a third, Morrisonville, was relocated to make room for a plant.

Despite the accommodations made by individual communities, the committee came to the conclusion that the 85-mile stretch of Cancer Alley had, indeed, earned its name. It reported to the U.S. Commission on Civil Rights that "black communities in the corridor between Baton Rouge and New Orleans are disproportionately impacted by the present State and local government systems for permitting and expansion of hazardous waste and chemical facilities. These communities are most often located in rural and unincorporated areas, and residents are of low socioeconomic status with limited political influence facilities" (Louisiana Advisory Committee 1993, 63). The committee further concluded that a major reason for the existence of Cancer Alley was that "state and local governments have failed to establish regulations or safeguards to ensure such communities are reasonably protected from a high concentration of hazardous waste and industrial facilities and risks associated with living in and around such facilities" (Louisiana Advisory Committee 1993, 63).

Pesticide Exposure

For more than half a century, U.S. farmers have relied heavily on the use of synthetic chemicals to protect their crops from destruction by insects, rodents, microorganisms, and other predators. Names such as DDT, aldrin, malathion, parathion, chlordane, heptachlor, toxaphene, and Sevin are now familiar to many Americans whether they are engaged in agriculture or not. These chemicals have been popular with farmers because they kill pests effectively at relatively modest costs.

As the use of synthetic chemicals spread through agriculture, however, some important disadvantages became evident. Among other problems, some of the most popular pesticides proved to be toxic not only to pests but also to harmless or beneficial animals and to humans themselves. One of the most influential books on the problems posed by pesticides was Rachel Carson's *Silent Spring*, published in 1962. In her book, Carson wrote about a spring in which no bird songs could be heard (the "silent spring") because they had been killed by pesticides.

Responding to growing public concerns about the risks posed by pesticides, the EPA began regulating agricultural chemicals much more closely in the early 1970s. In 1972, for example, it banned the widely popular pesticide dichlorodiphenyl-trichloroethane (DDT). In the same year, the U.S. Congress passed amendments to the Federal Insecticide, Fungicide, and Rodenticide Act (FIFRA), imposing much stronger regulations on the use of many pesticides. With actions such as these, many Americans began to feel that the most serious environmental problems posed by synthetic pesticides were being brought under control.

However, pesticides continued to pose dangers to farmworkers. Many of the regulations passed to protect consumers from pesticides made no mention of the dangers posed to men and women working in the fields. Among the products banned by the EPA were a number of chlorinated hydrocarbon pesticides, such as aldrin, dieldrin, endrin, heptachlor, and chlordane. This class of pesticides tends to be persistent, meaning that they remain on fruits and vegetables long enough that they may be present when consumed by the general public. Farmers switched instead to pesticides known as organophosphates (malathion, parathion, and methyl parathion, for example) and carbamates (Sevin, Zectran, and Temik, for example). These compounds are more toxic but less persistent. They pose less threat to consumers (because they have degraded by the time fruits and vegetables reach the marketplace) but greater threat to farmworkers (Miller 2003).

The problem of pesticide exposure is an important issue in the environmental justice movement because a large proportion of farmworkers are Hispanic Americans, especially Mexican Americans. Many are illegal aliens who work for less-than-minimum wages, often under difficult and illegal working conditions. They are seldom in a position to complain about working conditions since objections are likely to be met with dismissal from their jobs.

It is true that laws have been passed to protect farmworkers. In general, however, those laws have been weak, poorly enforced, and, therefore, largely ineffective. For example, it is illegal to use pesticides on a field in such a way that workers will be sprayed directly. But this regulation does not necessarily protect workers on an adjacent field from being exposed to pesticides blown onto them. Also, state officials are allowed to use otherwise unlicensed pesticides in an emergency, such as the unexpected appearance of a pest. In such a case, the implication is that it is acceptable to expose farmworkers to a chemical hazard to save crops.

Some regulations exist governing the period of time after spraying during which workers are not allowed to reenter fields. Given the paucity of data on safe intervals for most pesticides, however, such regulations may not be effective (Wasserstrom and Wiles 1985; Moses 1989, 1993). Overall, one observer has concluded that "the reality [of existing pesticide legislation] is that the extant legislation leaves farmworkers virtually unprotected against pesticide hazards" (Perfecto 1992, 216).

Consumption of Toxic Fish from the Detroit River

Environmental inequities can exist in many forms. One that is probably less apparent than the cases described previously is in the dietary patterns of minorities who rely on fish taken from the Detroit River. An analysis of this issue was carried out by Patrick C. West, associate professor in the School of Natural Resources and Environment at the University of Michigan, and his colleagues in 1992. The study was motivated by the fact that state and federal guidelines currently specify the amount of toxic contaminants that may be released into surface waters. The regulated level of contaminants is determined, in turn, by assuming that the average person will consume a certain amount of fish per day. For example, the state of Michigan's Rule 1057, which regulates the discharge of toxic chemicals into the state's surface water, assumes that the average level of fish consumption in Michigan is 6.5 grams per person per day. Under those circumstances, an allowable amount of no more than 0.1 milligrams of a toxic substance in the fish would avoid endangering the health of the consumer.

The flaw in this logic, however, is that some residents of Michigan are likely to consume more than 6.5 grams of fish each day on a regular basis. A person who regularly eats 10 or 20 grams of fish, for example, would then be exposed to two or three times the toxic amount assumed by the regulatory agency. Rather than being protected by federal or state regulation, consumers are instead lulled into a false sense of safety.

West and his colleagues attempted to find out the extent to which this situation might exist in the state of Michigan and, if it did, the extent to which certain individuals might be exposed to unexpected environmental risks from eating fish. West's team made two interesting discoveries. First, they found that, among those who fish in the Detroit River, 21.7 percent of whites in the sample were more likely to fish for both recreational purposes and to obtain food for meals, whereas 58 percent of nonwhites fished for both recreational purposes and to obtain food.

In addition, the West research team found that members of all minority groups studied (African Americans, Native Americans, and other minorities) consumed about five times the amount of fish per day as assumed by Michigan's Rule 1057. The team concluded that the regulation was, therefore, inadequate to protect those who fish primarily for food rather than for recreation. They ended their report with the observation that "to rely on a policy of fish consumption advisories creates more of a hardship for minorities because a needed protein source is at stake, not just a chance to catch that big one that didn't get away" (West et al. 1992, 111).

Proposed Select Steel Corporation Plant (Flint, Michigan)

The "justice" of the environmental justice movement sometimes seems so obvious that one wonders how critics can complain about efforts to ameliorate cases of environmental inequities. One answer to that question can be found in a complex case involving plans by the Select Steel Corporation of America in the late 1990s to build a new factory in Michigan's Genesee County. Most residents of the county were delighted to hear of the plans as the new plant promised to bring with it at least 200 new jobs in an economically depressed region of the state. At least one group of citizens was not pleased with the proposal, however. Father Phil Schmitter and Sister Joanne Chiaverini of the St. Francis Prayer Center in Flint, Genesee County's largest city, raised objections with the EPA, claiming that pollution from the new plant would cause disproportionate health effects on minority people living near the construction site.

The complaint was the first of its kind to be presented to the EPA since the agency had adopted its *Interim Guidance for Investigating Title VI Administrative Complaints Challenging Permits,* a set of guidelines for ruling on environmental justice cases. The EPA soon discovered, however, that its guidelines were too vague to permit a clear decision on some facets of the Select Steel case. It eventually decided, however, that there was no evidence that the plant would produce toxic releases greater than standards set by the state itself and by the EPA. It rejected the Schmitter-Chiaverini claim.

Dissatisfied with this result, complainants asked the EPA to reconsider its decision. Meanwhile, Select Steel was eager to get a resolution to its problems. It threatened to move to Ohio if approval were not granted for the Genesee plant, and, finally, in April 1999, it announced that it would be building the new plant in Delta Township, near Lansing, Michigan. At this point, very few people were satisfied with the outcome of this environmental justice complaint, some feeling that justice had not been done, and others viewing the complaint as an overreaction by environmentalists to a nonproblem. As Michigan governor John Engler said about the case, "The EPA is imposing their bureaucratic will over this community and punishing a company with the latest environmental standards, all because of a baseless complaint. . . . The net result is that the EPA is a job killer" (Alex J. Sagady & Associates 1998).

Sweet Valley and Cobb Town, Alabama

Residents of the Sweet Valley and Cobb Town neighborhoods of Anniston, Alabama, generally expected to live out their lives in their poor but comfortable homes. Then they learned about the toxic wastes produced by the Monsanto PCB (polychlorinated biphenyl) plant only a few hundred feet from their houses. Monsanto had purchased the plant in the 1930s and taken over production of PCBs from its previous owners. Polychlorinated biphenyls are a class of organic compounds once widely used in a variety of products, including coatings of electrical wiring and electronic components, flame retardants, sealants, adhesives, paints, coolants and insulating fluids for transformers, additives in pesticides, and carbonless copy paper. Both Monsanto and the African Americans who called Sweet Valley and Cobb Town home shared the lowest land in Anniston for the same reason: it was inexpensive. Wealthier white residents generally lived in higher parts of the town, less exposed to toxic gases and runoff from the plant.

By the 1970s, almost everyone had become aware of the harmful effects of PCBs on the environment. The EPA eventually banned their production and use in 1979 (although Monsanto had stopped production in Anniston eight years earlier). Still, it was not until the early 1990s that Sweet Valley and Cobb Town residents realized the potential health threats they faced from exposure to PCBs. They began to see what they thought of as an unusually large number of cases of cancer, birth defects, and other health problems that they attributed to years of PCB exposure. Organized as the Sweet Valley/Cobb Town Environmental Justice Task Force, they filed suit and sought EPA assistance in gaining compensation from Monsanto for their health problems.

Six years later, the case was resolved. In 2001, Monsanto agreed to pay residents of the area $42.8 million to facilitate their relocation. Even though valid and reliable evidence for a connection between PCBs and human health disorders was still largely absent, the company decided that it no longer wanted to fight what appeared to be a losing battle. Environmental justice for the citizens of Sweet Valley and Cobb Town had been achieved, but at what health cost still remained to be seen.

West Harlem Environmental Action and Diesel Buses

For low-income and minority communities, examples of environmental racism are often ubiquitous, if not immediately obvious. Over the last decade, the disproportionate effect of highway construction and public transit systems on such communities has become increasingly evident and of greater concern to environmental justice groups. One of the first instances in which a complaint was filed about environmental inequities in transportation occurred in 2000, when West Harlem Environmental Action (WE ACT) filed a complaint with the EPA. That complaint claimed that New York City's Metropolitan Transit Authority (MTA) disproportionately affected citizens in its part of the city. WE ACT pointed out that six of the city's eight major bus depots are located in Harlem. Diesel buses traveling throughout the city almost all have to drive through Harlem at some point in their trips. The pollutants released by these buses, WE ACT said, are an important factor in the abnormally high rate of asthma among children in the region.

WE ACT pointed out in its complaint that options were available for reducing diesel bus emissions. For example, the MTA's plans to close a downtown terminal could be reversed, distributing pollution problems to a wider part of the city. Also, the agency could consider using buses that produce less pollution, such as natural gas or diesel-electric hybrids. In response to the organization's criticisms, the MTA agreed to take some steps to deal with WE ACT's concerns, primarily the use of less-polluting vehicles.

In 2005, the EPA formally rejected the WE ACT complaint, explaining that it had found no evidence of intentional environmental racism in the MTA's policies and practices. It did admit, however, that some of the concerns expressed by the organization in its formal filing did have justification. Meanwhile, the MTA continued to waffle on its agreement with WEACT about dealing with the pollution caused by its buses in Harlem. In 2004, it announced that it was no longer considering shifting to natural gas–powered buses because diesel technology had improved so much that MTA vehicles no longer posed a health threat to residents of Harlem. Four years later, it changed its mind again, announcing that it would begin buying fuel-efficient, low-polluting diesel-electric hybrid buses for use in the city.

But the battle over buses was still not over. Reacting to the MTA's 2004 decision, WE ACT executive director Peggy Shepard said that "we do understand that C.N.G. [compressed natural gas] and hybrid diesel buses are measuring about the same level in terms of emissions. But without an agreement in place on a variety of environmental measures that we want them to implement in Harlem, I cannot say we are not opposed" (Chan 2005).

Southern Nuclear Operating Company

Proponents of environmental justice take pride in pointing to the many cases in which their complaints have been successful and some measure of recompense has been provided to victims of environmental inequities. A broader view of the topic reveals, however, that activists frequently lose such cases. Understanding the reasons for these losses is important because it helps environmental justice advocates deal with their problems more effectively in the future. An example of such a case is the petition filed in 2007 by five groups interested in environmental justice: the Center for a Sustainable Coast, Savannah Riverkeeper, the Southern Alliance for Clean Energy, the Atlanta Women's Action for New Directions, and the Blue Ridge Environmental Defense League.

These groups appeared before the Nuclear Regulatory Commission's Atomic Safety and Licensing Board Panel to challenge Southern Nuclear Operating (SNO) Company's petition for an early site permit at its Vogtle Electric Generating Plant near Waynesboro, Georgia. The five groups argued that the SNO environmental report had not given adequate attention to the possible disproportionate health effects produced by the plant on low-income and minority populations in its vicinity. They pointed out that residents of the area already have unusually high cancer rates (as a result of eating fish contaminated with toxic chemicals) and that the plant's operation may only magnify this effect. Finally, they said that SNO had made no special provisions for low-income and minority groups in case of a nuclear accident at the plant.

In March 2007, the panel ruled on the group's complaint. That ruling reflected a theme common among cases in which environmental justice activists have failed to make their arguments before governmental agencies. The panel said that the groups had failed to meet two essential requirements established for deciding environmental justice cases: (1) providing evidence for "the alleged existence of adverse impacts or harm on the physical or human environment" and (2) showing that "these purported adverse impacts could disproportionately affect poor or minority communities in the vicinity of the facility at issue." For these reasons, the panel rejected the groups' complaint (Lee 2007).

Environmental Justice: Environmentalism and Civil Rights

The movement for environmental justice represents the confluence of two older movements in the United States—the environmentalist and civil rights movements. The origins of the environmentalist movement in this country can be traced to the late 19th century. Before then, concern about the environment was virtually nonexistent in the United States. Instead, the dominant philosophy was that of the biblical injunction to "go forth and conquer the Earth." Early pioneers sweeping westward from the original 13 colonies encountered unimagined riches of land, minerals, timber, and wildlife. They harvested and used those resources without consideration about preserving or conserving them for some future use. Indeed, official governmental policy encouraged this rape of the land and its resources. The Preemption Act of 1841, for example, allowed a settler and his family to buy homestead land of up to 480 acres for 50 cents an acre. In many cases, lumber companies hired gangs of men to claim the land and hand it over to the companies. A similar practice was followed by cattle barons, who hired cowboys to claim homesteading lands under the Desert Land Act of 1877 and then use the lands for grazing their cattle (Petulla 1988).

It was only with the end of the westward migration, as pioneers reached the Pacific Coast, that some individuals began to realize that the continent's resources were not, in fact, unlimited. This realization spurred the first environmental movement in the United States. That movement actually consisted of two themes—conservationist and preservationist. The term *conservation* refers to the practice of harvesting natural resources in a controlled manner so that they will be available for future generations. The current U.S. government policies of "multiple use, sustained yield" reflect this philosophy. Conservationists do not argue that nature is sacred and inviolable but that it provides humans with many valuable resources that they must learn to use wisely. One of the most articulate spokespersons for the early conservationist movement was Gifford Pinchot, chief of the Division of Forestry (later the National Forest Service) in the Department of Agriculture (Wild 1979).

In contrast to the views of the conservationists was the philosophy of preservation. Preservationists argue that there is a beauty and value in nature that has nothing to do with the commercial value for humans. Humans have the obligation to protect vast portions of the natural world for no other reason than for its inherent value. According to preservationists, large segments of the world's forests, deserts, grasslands, and other natural resources should be set aside forever, for their own protection and for their enjoyment by humans. The philosophy of preservation owes much to the early writings of Henry David Thoreau. Thoreau once wrote that "there is a subtle magnetism in Nature, which, if we unconsciously yield to it, will direct us aright" (Thoreau 1962, 265). Perhaps the most famous preservationist was John Muir, one of the founders of the Sierra Club and the man largely responsible for convincing President Theodore Roosevelt of the need for the U.S. National Park system.

The traditional environmental movement, then, was one that strongly emphasized the connection between humans and nature. Hiking, backpacking, mountain climbing, bird-watching, photography, canoeing, and many other forms of recreation were the preferred activities of those who called themselves "environmentalists" until the mid-20th century. The environmental groups that sprung up within this movement—groups such as the Sierra Club, the Wilderness Society, and the National Parks and Conservation Association—strongly reflected that emphasis. Their goal was to protect the natural environment for a variety of human uses, whether for pure enjoyment or limited commercial exploitation.

The Modern Environmental Movement

The 1960s saw the rise of a quite different kind of environmental movement. This movement was inspired by a dawning recognition of the havoc humans were wreaking on the natural environment. Some people credit Rachel Carson's book *Silent Spring* (1962) as the defining moment in the modern environmental movement. Carson showed how the indiscriminate use of pesticides killed insects and the birds who fed on them. She called attention to the fact that many of the wonders of modern science and technology that made life healthier, safer, and more pleasant for humans were, at the same time, causing extensive damage to the physical and biological environment.

Out of this dawning realization grew a host of new environmental groups, groups whose focus was not primarily on the enjoyment of nature but on issues such as air and water pollution, hazardous waste disposal, and land use. Examples of such groups are the Acid Rain Foundation, the Citizens Clearinghouse for Hazardous Wastes, the Environmental Action Foundation, and the Friends of the Earth.

Groups that make up the traditional and modern environmental movements are often considered mainstream environmentalists. Early activists in the environmental justice movement have often argued, however, that they have little in common with mainstream environmentalism. One reason for this position is that mainstream environmentalist groups tend to be largely white and middle or upper class. One study of the membership of mainstream environmental groups in the 1940s, for example, found that 96 percent of the 1,468 respondents classified themselves as caucasian/European. Almost half (48 percent) had a total family income of more than $10,000, and 15 percent had a total family income of more than $25,000. In comparison, only

10 percent of American families made $13,000 or more at the time of the survey (as cited in Smith 1974).

In addition, the evidence suggests that most mainstream environmentalist groups have traditionally had little interest in issues faced by poor, minority, urban people. When members of the Sierra Club were asked in the 1970s, for example, whether the club should concern itself with the conservation problems of such special groups as the urban poor and ethnic minorities, about 40 percent said "no" and about 15 percent said "yes" strongly (as cited in Smith 1974). When a similar poll was conducted more than a decade later, a proposal to increase involvement in environmental issues faced by the urban poor and communities of color was defeated by a vote of about three to one (Freudenberg 1984).

Under the circumstances, it is hardly surprising that at least some observers have had harsh words for mainstream environmentalism. For example, Richard Hatcher, then mayor of the city of Gary, Indiana, observed in 1970 that "the nation's concern with the environment has done what George Wallace had been unable to do: distract the nation from the human problems of black and brown Americans" ("The Rise of Anti-Ecology?" 1970, 42).

In fact, the accomplishments of mainstream environmentalism have sometimes exacerbated the environmental problems of low-income people and people of color. As an example, new regulations on the use of pesticides in agriculture—rules enacted under the pressure of mainstream environmental groups—have in some cases increased the risk to farmworkers. Agricultural business people have replaced pesticides of lower toxicity and longer persistence (thereby providing greater protection to consumers) with pesticides of greater toxicity and shorter persistence (thus increasing the risk to farmworkers) (Perfecto 1992).

In any case, many of those active in the environmental justice movement retain doubts about the ability of mainstream environmentalist groups to understand or respond to the issues about which people of color are most concerned. In assessing the current relationship between mainstream environmentalism and the environmental justice movement, Robert Bullard has written that "the mainstream environmental movement has proven that it can help enhance the quality of life in this country. . . . Yet, few of these groups have actively involved themselves in environmental conflicts involving communities of color. Because of this, it's unlikely that we will see a mass influx of people of color into the national environmental groups any time soon. A continuing growth in their own grassroots organizations is more likely" (1993, 38).

The Civil Rights Movement

For some historians, the modern civil rights movement in the United States can be traced to December 1, 1955, when African American seamstress Rosa Parks refused to give up her seat in the first row of the "colored" section of a Montgomery, Alabama, bus to a white man. The event precipitated a yearlong boycott of Montgomery buses by African Americans in the city and culminated in a decision by the U.S. Supreme Court that segregation on public bus systems was illegal.

The next decade was marked by the battle by African Americans and their allies to eliminate the whole fabric of segregation that had been constructed in the South over the preceding century. Important events included the Alabama Freedom Rides of 1961; civil rights demonstrations in Birmingham, Alabama, and the March on Washington, D.C., in 1963; and the marches on Selma, Alabama, in 1965. Although protests were strongly influenced by the philosophy of nonviolence, they were often met with extreme violence, resulting in injuries to and the deaths of hundreds of protestors.

The civil rights movement's major goals were slowly accomplished, however, with the passage of the Voting Rights Act of 1957, the Civil Rights Act of 1964, the Fair Housing Act of 1968, and similar legislation. Courts also confirmed that segregation of any kind in American society was unconstitutional and set forth rigorous programs for the elimination of bias in education, housing, employment, and other fields. Although the centuries-old evils of segregation were hardly eliminated, the nation's highest legislative and judicial bodies had made clear that official forms of apartheid in the United States were illegal (Grossman 1993; Morris 1984).

The environmental justice movement probably owes more to one parent—the civil rights movement—than to the other— the environmental movement. Although the issues with which environmental justice deals are environmental issues—air and water pollution, hazardous waste sites, and the like—the political philosophy that underlies action is a reflection of the civil rights philosophy. In addition, most of the groups currently acting on environmental justice issues began their existence as civil rights groups and not environmental organizations. As Robert Bullard has said, "The push for environmental equity is an extension of the civil rights movement, a movement in which direct confrontation and the politics of protest have been real weapons" (1994, 13).

Environmental Justice Comes of Age

Walter E. Fauntroy, then representative from the District of Columbia to the U.S. House of Representatives, participated in and was arrested for his part in the 1982 Warren County protest described previously. Following his release from jail, Fauntroy asked the U.S. General Accounting Office (GAO) to conduct a study on the relationship between pollution and minority communities. The report found that three of the four largest landfills located in the Southeast were located in predominantly poor and African American communities (U.S. General Accounting Office 1983).

The GAO study was rather modest in its scope, but it was important for another reason. It brought to the attention of the general public, probably for the first time, a possible association between race and income on the one hand and exposure to environmental hazards on the other. In addition, it highlighted the serious lack of fundamental research on this issue.

The Commission on Racial Justice of the United Church of Christ

In response to the gap in fundamental research, the Commission on Racial Justice of the United Church of Christ decided to initiate an even more extensive study of the environmental problems faced by minority and poor communities. The commission analyzed the location of hazardous waste sites in all zip (postal) codes and counties in the United States. The results of the study were released in 1987 and were immediately hailed by experts in both environmental and civil rights communities. In a 1987 editorial, for example, the *Atlanta Constitution* observed that the study put an end to any doubts that white America had been pursuing a policy of dumping its garbage in black neighborhoods. The commission report, it said, clearly pointed out where public health officials and environmental inspectors needed to look for potential problems before they became health hazards.

What was it that the Commission on Racial Justice had actually found? Among the most important points in the commission report were the following:

- People of color are twice as likely to live in a community with a commercial hazardous waste site than are whites. They are three times as likely to live in a community with more than one hazardous waste site.
- About 60 percent of African Americans live in a community with an abandoned hazardous waste site.
- The average annual income of people living in communities with a hazardous waste site is significantly less than those living in communities without such a site.
- In spite of the preceding point, race, more than income, is a more reliable factor in predicting the presence of a hazardous waste site.
- Three of the five largest hazardous waste sites, accounting for about 40 percent of the nation's total landfill capacity, are located in communities that are predominantly African American or Latino.
- Fifteen million African Americans and 8 million Hispanic Americans live in communities with one or more hazardous waste sites.
- Of the six cities with the largest number of hazardous waste sites, African Americans make up a far greater fraction of residents than do whites. The numbers for those cities are as follows: Memphis (43.3 percent African Americans; 173 sites), St. Louis (27.5 percent; 160 sites), Houston (23.6 percent; 152 sites), Cleveland (23.7 percent; 106 sites), Chicago (37.2 percent; 103 sites), and Atlanta (46.1 percent; 94 sites). For comparison, African Americans make up 11.7 percent of the general U.S. population. (Commission on Racial Justice 1987)

As dramatic as the Commission on Racial Justice report was, it was certainly not the first serious study of the relationship among environmental hazards, race, and income. For example, an early study by David Harrison found that the costs of air pollution were more likely to be borne by low-income groups than by middle- or high-income groups (1975).

Another important study was the one completed by Robert D. Bullard and reported in 1983. Bullard found that the pattern of siting for municipal landfills, incinerators, waste transfer stations, and other forms of waste disposal facilities in Houston was similar to what was later reported by the Commission on Racial Justice. That is, all five of the city-owned landfills were located in predominantly African American neighborhoods; six of eight garbage incinerators were located in such neighborhoods, with the remaining two in a predominantly Hispanic American neighborhood and a predominantly white area (Bullard 1983). Over three decades, a number of studies similar to those of Bullard's have been conducted in specific cities, counties, and states. (The bibliography in Chapter 8 lists the reports from many of these studies.)

The Michigan Conference

As the decade of the 1990s dawned, environmental justice was a topic of considerable interest across a wide spectrum of citizens from all walks of life. In 1990, two academicians relatively recently introduced to the issues of environmental justice—Bunyan Bryant and Paul Mohai at the University of Michigan School of Natural Resources (now the School of Natural Resources and Environment)—convened a conference of scholars and activists in Ann Arbor to present their latest research, discuss their ideas, and consider possible solutions to the problems of environmental racism and environmental inequities. A number of papers presented at the Michigan Conference were later reprinted in *Race and the Incidence of Environmental Hazards,* edited by Bryant and Mohai.

The Michigan Conference was important because of other events it initiated. First, participants at the conference drafted a letter to Louis W. Sullivan, secretary of the U.S. Department of Health and Human Services; William K. Reilly, administrator of the EPA; and Michael R. Deland, chair of the Council on Environmental Quality. The letter asked for an opportunity to meet with government officials to discuss a number of topics relating to environmental justice. Among these issues was the request for more research on the nature of environmental inequities in the United States, the development of projects targeted at providing aid to low-income and minority populations, the appointment of special assistants for environmental equity in government agencies, and the development of a policy statement on environmental equity. Representatives of the Michigan Conference met with Reilly and other government officials on September 13, 1990, in what appeared to be a highly productive meeting.

Governmental Recognition in the 1990s

Although the administration of President George H. W. Bush took no concrete action to deal with the issues of environmental justice, an important seed had been planted. That seed came to fruition with the election of Bill Clinton as president in 1992. Clinton appointed Benjamin Chavis and Robert Bullard, both activists in the field of environmental justice, to serve on his transition team in the area of natural resources (which included the EPA and the Departments of the Interior, Energy, and Agriculture). Then, on February 11, 1994, President Clinton issued Executive Order 12898 establishing the Office of Environmental Justice within the EPA and the National Environmental Justice Advisory Council (NEJAC).

Clinton's action is arguably the most important step to occur in the environmental justice movement in its short history. The creation of the Office of Environmental Justice and the NEJAC provided the mechanism and the financial support needed for those concerned about environmental justice issues to meet at regular intervals, design programs of research, present their concerns directly to the federal administration, hold public meetings at which groups and individuals can testify, and comment on federal programs that have a relevance to environmental inequities.

The NEJAC consists of about two dozen members selected from community-based groups, business and industry, academia, tribal governments, nongovernmental organizations, and environmental groups. The NEJAC is organized into four subcommittees—Waste and Facility Siting, Enforcement, Health and Research, and Public Participation and Accountability.

First National People of Color Environmental Leadership Summit

Another important event in the organization of a national environmental justice movement took place in October 1991. Under the auspices of the Commission on Racial Justice of the United Church of Christ, the First National People of Color Environmental Leadership Summit was convened in Washington, D.C. More than 500 participants from a wide variety of ethnic, racial, cultural, and economic groups came together to talk about their mutual concerns about the environment and environmental inequities. More than 50 presentations, workshops, panels, caucuses, and other sessions were held during the three-day conference. At the conference's conclusion, the participants adopted a 17-part statement, "Principles of Environmental Justice," outlining their combined view on the state of environmental inequities in the United States. The principles have become a guideline for much of the thought and work that now takes place in environmental justice groups throughout the nation. (See Chapter 6 for a statement of the principles.)

Environmental Justice as a Social Movement

The environmental justice movement in the United States stands out among other social movements. Although it has begun to take on some of the trappings familiar to the movements from which it sprang, namely, a centralized organization located in large governmental and nongovernmental bodies, it still remains to a large degree a highly decentralized movement. While academicians and activists were coming together at the Michigan Conference and in the development of the EPA Office of Environmental Justice, dozens of small, local groups were being organized to deal with the very real issues faced by individual neighborhoods and communities. (For a superb discussion of all facets of the environmental justice movement, see Lazarus [1993].)

The 1990s saw the formation, for example, of groups such as the Coalition for Environmental Consciousness in Ridgeville, Alabama; California Indians for Cultural and Environmental Protection in Santa Ysabel, California; the Citizens League Opposed to Unwanted Toxins in Tifton, Georgia; Ke Kua'aina Hanauna Hou in Kaunakakai, Hawaii; the Flint-Genesee United for Action, Justice, and Environmental Safety in Flint, Michigan; Concerned Citizens of Sunland Park in Sunland Park, New Mexico; Fort Greene Community Action Network in Brooklyn, New York; Eufaula Street Landfill Committee in Fayetteville, North

Carolina; Environmental Services Office of the Cherokee Nation in Tahlequah, Oklahoma; People Organized in Defense of Earth and Its Resources in Austin, Texas; and the Environmental Center for New Canadians in Toronto.

Representatives of such groups often stay in contact with each other and draw on each other's resources through tribal, regional, cultural, or other affiliations, such as the Southwest Network for Environmental and Economic Justice, the Asian Pacific Environment Network, the Gulf Coast Tenants Organization, or the National Congress of American Indians. Yet, in many cases, the bulk of their actual work occurs within a neighborhood, a small town, a tribe or reservation, or a county.

Local groups also stay in contact with each other through a number of journals, magazines, newsletters, and other publications that have grown up over the past three decades. These include *Race, Poverty, and the Environment* (published by Urban Habitat), *Everyone's Backyard* (Center for Health, Environment, and Justice), *RACHEL's Hazardous Waste News* (Environmental Research Foundation), *Toxic Times* (Chemical Injury Information Network), *Voces Unidas* (Southwest Organizing Project), and *PAN* magazine (Pesticide Action Network North America).

An interesting feature of the modern environmental justice movement is the critical role played by women. Mothers, housewives, secretaries, and rural women have long been at the forefront of protests, community action, leafleting, public testimony, and other actions on topics concerning environmental inequities and environmental racism. A number of writers have observed that many issues in the field of environmental justice strike close to the individual home and thus arouse more interest among housewives and mothers than would issues from mainstream environmentalism, such as the protection of endangered species or the development of national parks.

Celene Krauss, assistant professor of sociology and coordinator of Women's Studies at Kean College of New Jersey, for example, has drawn on interviews with women active in the environmental justice movement to place their participation in a broader context of feminist political philosophy. She concludes that women who can be described as "female blue-collar activists" often have a political philosophy similar to those in the field of environmental justice and a suspicion of the relevance of mainstream environmental organizations to the concerns of women like themselves (Krauss 1994).

Growth and Challenges

By the first decade of the 21st century, the environmental justice movement had become a strong, diversified, active, and successful crusade in the United States. Each new edition of the basic reference text, the *People of Color Environmental Groups Directory*, showed a significant increase in the number of environmental groups in the United States, Puerto Rico, Canada, and Mexico: from 205 such groups in 1992 to about 300 groups in 1994 to more than 400 groups in 2000. And stories of successful battles by local environmental groups against a variety of corporate groups and governmental agencies continued to flood the academic and movement literature (see the chronology in Chapter 4 and resources in Chapter 8 for more detail).

By the mid-1990s, environmental justice groups had found that they had two powerful tools for exposing and acting against cases of environmental injustice: President Clinton's Executive Order 12898 and Title VI of the Civil Rights Act of 1964. Title VI specifies that "no person in the United States shall, on the ground of race, color, or national origin, be excluded from participation in, be denied the benefits of, or be subjected to discrimination under any program or activity receiving Federal financial assistance" (Title VI of the 1964 Civil Rights Act). Since a great many governmental and nongovernmental organizations and agencies receive at least some part of their budget from the federal government, these organizations and agencies are subject to the provisions of Title VI of the Civil Rights Act. Groups concerned about environmental justice realized that they could file complaints with the EPA against organizations and agencies that they believed practiced environmental racism.

The first case on record involving this type of action took place in 1992, when members of the St. Francis Prayer Center in Flint, Michigan, filed a complaint with the EPA to prevent construction of a new "mini-mill" plant near Flint by the aforementioned Select Steel Corporation of America. (Because the EPA apparently mislaid the complaint, it was not acted upon until 1998; U.S. EPA 1998.) A year later, a group of environmental justice organizations also filed complaints with the EPA against the Louisiana and Mississippi Departments of Environmental Quality, claiming that, in violation of the provisions of Title VI, they discriminated against low-income and minority communities in making environmental policies and decisions. Within a matter of years, the instances of using Title VI as a tool against industries and government agencies had become a flood, with well over 100 cases being filed by the year 2000.

Environmental justice groups have not, however, met with a great deal of success thus far in the use of Title VI as a means of gaining redress against situations that they regard as environmental racism. An article published in the November 2003 issue of the *New York Law Journal* found that, of the 136 claims submitted by environmental justice groups based on Title VI standards to date, all decisions had been to deny the claim. Of the 136 cases, 75 were rejected without investigation, 26 were investigated and found to have no evidence of discrimination, and 30 were still pending at the time of the article. The most common reason for rejecting a claim, the article reported, was that the complaint had not been filed in a timely manner. The EPA has also been perhaps unduly deliberate in considering complaints presented to it. In one case, it issued a decision almost 10 years after the original complaint had been filed (Gerrard 2003, 3). A summary of similar cases that had been presented to the EPA since the law journal article was published can be found on the American Bar Association's Update Web site. More than 20 of the cases discussed resulted in rejection of claims by the EPA; less than 5 were favorably reviewed or accepted by the agency or by any other state or federal court or agency (American Bar Association 2008).

President Bill Clinton's Executive Order 12898 on environmental justice has also had somewhat less of an impact on environmental inequities than many people had once hoped and expected. A major problem appears to have been the low priority placed on environmental justice issues by the administration of President George W. Bush. Over the eight-year period of Bush's administration, a number of governmental and nongovernmental commissions studied the implementation of Executive Order 12898. For example, the inspector general of the EPA issued a report in March 2004 noting that little progress had been made in implementing the executive order and that, in fact, the Bush administration had reinterpreted the meaning of environmental justice to exclude low-income and minority populations (Office of Inspector General 2004).

Later the same year, the U.S. Commission on Civil Rights issued a comprehensive review of the status of civil rights issues during Bush's first term. The commission concluded that despite a decade of active pursuit, Executive Order 12898 is still not part of EPA's core mission. President Bush did not consider implementation of the order a primary goal. . . . [The] EPA has not developed a clear vision or a comprehensive strategic plan, nor established values, goals, expectations, and performance measurements regarding the order. Further, EPA has not provided regional or program offices with standards for what constitutes a minority or low-income community, or defined the term "disproportionately" as it relates to environmental justice. If EPA does not identify parameters for environmental justice, it will not be able to comply with Executive Order 12898. (U.S. Commission on Civil Rights 2004, 75–76)

As of late 2008, a comparable scorecard on the Bush administration's record during its second four-year term was not available. It can be noted, however, that the Republican Party did not include any mention of environmental justice in its 2004 platform (the Democratic Party platform did have a section under this title), and little progress appears to have been made in the implementation of Title VI or Executive Order 12898 policies.

The record of the U.S. Congress in dealing with environmental justice issues has not been much better than that of the administration. Rep. John Lewis (D-GA) and Sen. Al Gore Jr. (D-TN) introduced an environmental justice act into Congress in 1992, but the bill was not acted upon. A year later, Lewis and Sen. Max Baucus (D-MT) reintroduced the same act, again without success. Since that time, some version of the Lewis-Gore-Baucus bill has been introduced into Congress in 1998, 1999, 2002, 2003, 2005, 2007, and 2008. Since 1998, a primary purpose of the environmental justice legislation has been to codify Executive Order 12898 to provide it with greater legal force. In no case has an environmental justice bill come even close to passage; such bills often do not get out of the committees to which they are assigned.

In spite of the somewhat dismal record of the federal government in the area of environmental justice over the past decade, a number of individual success stories can be told. These stories often involve successful court cases at the local and state level and, even more often, successful negotiations between polluting industries and local grassroots environmental justice organizations. A typical example is the ongoing battle by residents of the Baldwin Hills neighborhood of Los Angeles to prevent despoliation of a large open space in their district. The open space in question consists of about two square miles of undeveloped land, the largest such area in urban Los Angeles.

In 2001, the state of California announced plans to grant a license to La Jolla Energy Development, Inc., to build a new power plant in the park as a way of responding to the state's energy crisis. With only 30 days to prepare for a hearing on the proposal, residents of Baldwin Hills (78.5 percent African American in the 2000 census) quickly brought together a coalition of concerned citizens and organizations to object to the plan. The coalition was able to mass more than 1,000 people at the June 21 hearing on the proposal. Faced with such daunting opposition, La Jolla Energy eventually withdrew its plans to build in Baldwin Hills park.

That victory was somewhat short-lived, however, as the city of Los Angeles announced only two years later that it planned to build a garbage dump on the fringes of Baldwin Hills park. Again, residents came together to oppose those plans and, again, they won. The city decided to place the "solid waste transfer station" in another location. Finally (if it truly turns out to be "finally"), residents of Baldwin Hills were able to convince the city in 2007 to place a moratorium on the construction of 24 new oil wells in the park (although wells had been in place in some parts of the park since the 1920s). Whether the 2007 decision marks the end of attempts to encroach on the park is still to be seen.

The modern environmental justice movement in the United States poses something of an enigma for those concerned about environmental and civil rights issues. On the one hand, the movement has achieved some amazing successes in its less-than-three-decade history. Poor people and minority communities have won redress and compensation from giant multinational corporations for the health risks they pose. Many states, regions, and local governments have acknowledged the significance of environmental justice issues and have passed laws and adopted regulations to ameliorate the effects of environmental inequities.

But much remains to be done. The federal government, after an impressive early initiative on behalf of environmental rights in the 1990s, has largely ignored the issue since. The president, Congress, and the courts at nearly all levels appear not to recognize the existence of environmental inequities or their need to act against such problems. What stumbling blocks do environmental justice activists still face, what basic questions about environmental inequities remain, and what can be done to more fully reach the goals of environmental equity that were first highlighted in the 1982 protest in Warren County? Chapter 2 examines these issues.

References

Alex J. Sagady & Associates. 1998. "E-M:/Engler News Release on Select Steel Case." [Online article or information; retrieved 5/14/08.] http://www.great-lakes.net/lists/enviro-mich/.

American Bar Association. Section of Environment, Energy, and Resources. 2008. "The Law of Environmental Justice: Update Service." [Online article or information; retrieved 4/22/08.] http://www.abanet.org/environ/committees/envtab/ejupdates.html.

Bullard, Robert D. 1983. "Solid Waste Sites and the Black Houston Community." *Sociological Inquiry* 53 (2–3): 273–288.

Bullard, Robert D. 1993. "Anatomy of Environmental Racism and the Environmental Justice Movement." In *Confronting Environmental Racism: Voices from the Grassroots,* edited by R. D. Bullard, chap. 1. Boston: South End Press.

Bullard, Robert D. 1994. *Unequal Protection: Environmental Justice and Communities of Color.* San Francisco: Sierra Club Books.

Bullard, Robert D. 2000. *Dumping in Dixie: Race, Class, and Environmental Quality,* 3rd ed. Boulder, CO: Westview Press.

Chan, Sewell. 2005 (March 28). "City to Buy Diesel-Electric Buses, Not Natural Gas Ones." *New York Times.* [Online article or information; retrieved 5/14/08.] http://www.nytimes.com/2005/03/28/nyregion/28bus.html?n=Top/Reference/Times%20Topics/Organizations/N/New%20York%20City%20Transit.

Commission on Racial Justice of the United Church of Christ. 1987. *Toxic Wastes and Race in the United States: A National Report on the Racial and Socio-Economic Characteristics of Communities Surrounding Hazardous Waste Sites.* New York: United Church of Christ.

Freudenberg, N. 1984. "Not in Our Backyards: Community Action for Health and the Environment." *Monthly Review Press* 22:34–39.

Gerrard, Michael B. 2003. "EPA Dismissal of Civil Rights Complaints." *New York Law Journal,* November 28, 3+.

Griffith, Matt, Mansoureh Tajik, and Steve Wing. 2007. "Patterns of Agricultural Pesticide Use in Relation to Socioeconomic Characteristics of the Population in the Rural U.S. South." *International Journal of Health Services* 37:259–277.

Grossman, Mark. 1993. *The ABC-CLIO Companion to the Civil Rights Movement.* Santa Barbara, CA: ABC-CLIO.

Harrison, David, Jr. 1975. *Who Pays for Clean Air: The Cost and Benefit Distribution of Automobile Emission Standards.* Cambridge, MA: Ballinger.

Krauss, Celene. 1994. "Women of Color on the Front Line." In *Unequal Protection: Environmental Justice and Communities of Color,* edited by Robert D. Bullard, chap. 14. San Francisco: Sierra Club Books.

Lazarus, Richard J. 1993. "Pursuing 'Environmental Justice': The Distributional Effects of Environmental Protection." *Northwestern University Law Review* 87 (March): 101–170.

Lee, Charles. 2007. "Environmental Justice in the News for the Week Ending April 20, 2007." [Online article or information; retrieved 4/20/07.] http://www.epa.gov/compliance/resources/newsletters/.

Louisiana Advisory Committee to the U.S. Commission on Civil Rights. 1993. *The Battle for Environmental Justice in Louisiana . . . Government, Industry, and the People.* Kansas City, MO: U.S. Commission on Civil Rights, Central Regional Office.

Maraniss, David, and Michael Weisskopf. 1989. "Jobs and Illness in Petrochemical Corridor." *Washington Post,* December 22, A1.

Miller, G. Tyler. 2003. *Living in the Environment: An Introduction to Environmental Science,* 13th ed. Belmont, CA: Wadsworth.

Morris, Aldon D. 1984. *The Origins of the Civil Rights Movement: Black Communities Organizing for Change.* New York: The Free Press.

Moses, Marion. 1989. "Pesticide Related Health Problems in Farm Workers." *American Association of Occupational Health Nurses Journal* 37:115–130.

Moses, Marion. 1993. "Farmworkers and Pesticides." In *Confronting Environmental Racism: Voices from the Grassroots,* edited by Robert D. Bullard, chap. 10. Boston: South End Press.

Office of Inspector General. U.S. Environmental Protection Agency. 2004. *EPA Needs to Consistently Implement the Intent of the Executive Order on Environmental Justice.* Washington, DC: Environmental Protection Agency.

Perfecto, Ivette. 1992. "Pesticide Exposure of Farm Workers and the International Connection." In *Race and the Incidence of Environmental Hazards: A Time for Discourse,* edited by Bunyan Bryant and Paul Mohai, chap. 14. Boulder, CO: Westview Press.

Petulla, Joseph M. 1988. *American Environmental History,* 2nd ed. San Francisco: Boyd & Fraser.

"The Rise of Anti-Ecology?" 1970. *Time,* August 3, p. 42.

Robinson, William Paul. 1992. "Uranium Production and Its Effects on Navajo Communities along the Rio Puerco in Western New Mexico." In *Race and the Incidence of Environmental Hazards: A Time for Discourse,* edited by Bunyan Bryant and Paul Mohai, chap. 12. Boulder, CO: Westview Press.

Smith, James Noel, ed. 1974. *Environmental Quality and Social Justice in Urban America.* Washington, DC: The Conservation Foundation.

Thoreau, Henry David. 1962. "Walking." In *Excursions.* New York: Corinth Books.

Title VI of the 1964 Civil Rights Act. [Online information; retrieved 5/5/08.] http://www.usdoj.gov/crt/cor/coord/titlevistat.htm. Updated November 13, 2000.

U.S. Commission on Civil Rights. Office of Civil Rights Evaluation. 2004. *Redefining Rights in America: The Civil Rights Record of the George W. Bush Administration, 2001–2004.* Washington, DC: U.S. Commission on Civil Rights.

U.S. Congress. 1993. House. Committee on the Judiciary. *Environmental Justice Hearings before the Subcommittee on Civil and Constitutional Rights.* 103rd Congress, 1st sess., March 3 and 4. Washington, DC: Government Printing Office.

U.S. Environmental Protection Agency (EPA). 1998. "In Re: Select Steel Corporation of America." [Online article or information; retrieved 5/15/08] http://www.gridauh.fr/sites/fr/fichier/447aec8d1257e.pdf.

U.S. General Accounting Office (GAO). 1983. *Siting of Hazardous Waste Landfills and Their Correlation with Racial and Economic Status of Surrounding Communities.* Washington, DC: Government Printing Office.

Wasserstrom, R. F., and R. Wiles. 1985. *Field Duty: U.S. Farmworkers and Pesticides Safety.* Study 3. Washington, DC: World Resources Institute, Center for Policy Research.

West, Patrick C., J. Mark Fly, Frances Larkin, and Robert W. Marans. 1992. "Minority Anglers and Toxic Fish Consumption: Evidence from a Statewide Survey of Michigan." In *Race and the Incidence of Environmental Hazards: A Time for Discourse,* edited by Bunyan Bryant and Paul Mohai, chap. 8. Boulder, CO: Westview Press.

Wild, Peter. 1979. *Pioneer Conservationists of Western America.* Missoula, MT: Mountain Press.

2

Problems, Controversies, and Solutions

The environmental justice movement has grown rapidly, has spread extensively throughout the world, and has achieved some remarkable successes in the past three decades. Today, most people who are interested in social justice, environmental issues, and human rights are aware of the context into which environmental justice has placed these topics for people of color and low-income communities.

Still, the movement has just begun to achieve the goals it has set out for itself and for society as a whole. Hazardous waste sites are still being sited in minority and low-income communities, petrochemical and other plants still deliver unacceptably high levels of pollutants to such communities, farmworkers and workers in hazardous industries are still exposed to health risks that could and should be reduced, and developed nations of the world still tend to think of developing nations as dumping grounds for products and wastes that they, themselves, do not allow in their own countries.

This chapter reviews some of the major problems and controversies still faced by the environmental justice movement. It considers whether environmental injustice really does exist and, if so, what evidence there is for it. It also asks to what extent residents of low-income and minority communities are exposed to inequitable amounts of environmental risks, whether these risks develop intentionally as the planned acts of government and/or industry, and what low-income and minority communities can do to combat these risks.

Do Environmental Inequities Actually Exist?

The fundamental premise of the environmental justice movement is that low-income and minority communities are exposed
to a disproportionate risk from hazardous waste sites, workplace hazards, polluting industries, and other facilities potentially harmful to human health. To some extent, this premise
was initially based on anecdotal evidence provided by people
living in such communities. By the late 1970s, however, research
studies also began to provide support for the premise. Perhaps
the best known of these early studies was one conducted by
Robert D. Bullard, then an assistant professor of sociology at
Texas Southern University in Houston. Bullard's research was
conducted to support a lawsuit filed by his wife, *Bean v. Southwestern Waste Management, Inc.*, in which plaintiffs claimed that
black citizens in Houston were being unfairly discriminated
against because of the siting of hazardous waste dumps in the
city. Bullard's research showed that all five city-owned garbage
dumps, six of the eight city-owned garbage incinerators, and
three of the four privately owned landfills were sited in black
neighborhoods. The evidence strongly supported claims of environmental inequities in the city. Bullard's research, completed
in 1979, was eventually published as "Solid Waste Sites and the
Black Houston Community" in the spring 1983 issue of *Sociological Inquiry.*

Over the next decade, researchers collected an increasingly
extensive body of evidence to support the premise that poor and
minority communities are disproportionately affected by hazardous working conditions and industrial facilities. These studies investigated problems such as the siting of radioactive waste
sites on Native American lands; the consumption of contaminated fish by minority communities around Puget Sound, Washington; the exposure of farmworkers to hazardous pesticides;
and the risks to minority communities of hazardous waste sites
in Massachusetts [for a summary of research studies conducted
on environmental inequities, see the U.S. Environmental Protection Agency's (2006) "Environmental Justice Bibliography Database"]. Many of the case studies presented in Chapter 1 were
documented during this period.

Perhaps the most significant of all studies on environmental
inequities conducted in the 1980s was sponsored by the Commission on Racial Justice of the United Church of Christ, discussed

in Chapter 1. The evidence from that study was so well documented and convincing that, by the time of its publication in 1987, many researchers had concluded that the case of environmental inequities in the United States had essentially been proved.

The United Church of Christ report was, however, by no means the only study on the existence of environmental inequities. As an additional example, Paul Mohai and Bunyan Bryant at the University of Michigan School of Natural Resources, organizers of the Michigan Conference on Environmental Justice, undertook a review in 1992 of 16 major studies on the relationship between race and income and the distribution of environmental hazards. They found significant correlations between these factors in all of the studies on which they reported (Mohai and Bryant 1992).

The conclusions drawn by Bullard, the United Church of Christ study group, Mohai, Bryant, and other researchers were not, however, entirely convincing to all investigators. By the early 1990s, a number of questions were being raised about the methodological accuracies of many early environmental justice studies. Some researchers had begun to ask if the apparent correlations between income and race on the one hand and environmental risk on the other were accurate or whether, in fact, they represented an artifact produced by faulty methodology. These investigators reported that the evidence from their own research seemed to contradict the basic premise of environmental justice; their studies seemed to suggest, that is, that low-income and minority communities are at no greater risk from hazardous exposure than are middle-income and nonminority communities.

Perhaps the most famous studies that came to these conclusions were those conducted by Douglas L. Anderton and his colleagues at the University of Massachusetts at Amherst in the early 1990s. The initial studies were funded by Chemical Waste Management, Inc., currently the largest waste management corporation in the world.

Anderton's research team focused on the smallest population units for which reliable national data are available. These units are the census tracts used by the U.S. Bureau of the Census in its decennial census. When the Anderton team compared the demographic characteristics of 1990 census tracts containing toxic waste treatment, storage, and disposal facilities (TSDFs) with similar tracts lacking such facilities, they found essentially no differences in the racial status of residents of the two tracts. They reported their results in a series of papers published in the mid- and late 1990s (Anderton et al. 1994a, 1994b; Anderton, Oakes, and Eagan 1996, 1997). In a summary of this research, Anderton's team drew three primary conclusions. First, the appearance of environmental inequity depends on the way in which areas of potential impact are defined. Second, if one uses census tract areas, as defined by the U.S. Census Bureau, hazardous waste sites are no more likely to be located in nonwhite neighborhoods as in other tracts. Third, the most consistently significant effect on the location of hazardous waste sites is the proportion of workers employed in industrial activities. That observation, the researchers said, should hardly be surprising since industrial operations tend to locate near other industrial facilities (Anderton et al. 1994a).

Anderton and his colleagues continued their research on environmental inequities based on race and income for a number of years. In 2000, he and colleague Pamela Davidson reported on a review of hazardous waste facilities governed by the Resource Conservation and Recovery Act (RCRA). Their results did not show disproportionate siting of such facilities based on income (facilities most commonly were placed in working-class neighborhoods), although the hazardous waste sites did tend to be located near minority communities. Overall, Davidson and Anderton concluded that "the siting of RCRA facilities does not merit high priority among the potential hazards and burdens to which minorities and the disadvantaged are disproportionately exposed" (Davidson and Anderton 2000, 466).

As might be expected, Anderton's research itself became the subject of some dispute, with other observers raising questions about the methodology used in his studies. Indeed, the scientific data on which claims of environmental inequities are based have been and continue to be the subject of considerable controversy among researchers. For example, some researchers point out the problems of proving that clear evidence exists for the connection between exposure to hazardous materials and cancer and other diseases, birth defects, and death. It can be very difficult to say that just because people live near a hazardous waste site, for example, their health problems are *caused* by exposure to those wastes.

Questions have also been raised about the way generalizations have been made by environmental justice advocates from specific situations in one locality to general patterns of environmental inequities throughout the nation or around the world. Situations are so different, some critics say, that each problem area must be studied and treated on its own.

The debate over the correlation among race, income, and environmental hazards is far from over. Scholarly articles in the first decade of the 21st century are still arguing the best approaches to use in studying these factors, and experts are still debating how extensive environmental inequities in the United States are. But other research questions remain also. To the extent that environmental inequities exist, researchers would like to know the answers to a number of related questions, such as how these conditions have arisen, whether they represent conscious intent on the part of industry and/or government, and how such inequities can be reduced in the future.

Precisely how environmental inequities develop is a complex question, one that has not yet been answered to everyone's complete satisfaction. Yet, some forces contributing to this state of affairs are reasonably clear. These forces can be characterized as follows:

- Progress and environmental degradation: Improvements in our way of life often lead to a diminished quality of the environment for all citizens, but often and especially for poor communities and communities of color.
- Siting decisions: The choices that governments and industries make in locating hazardous facilities can damage the environment.
- Access to tools of protest: Lack of education and experience can reduce the ability of some communities to understand and respond to environmental issues they face.
- Job blackmail: Individuals faced with a diminished environment are often forced to choose between keeping their jobs and accepting those conditions or battling for a better life and, perhaps, losing their jobs.
- Effects of environmental regulations: Ironically, governmental regulations designed to improve the overall environmental quality for an area sometimes have undesirable impacts on poor and minority communities.

Progress and Environmental Degradation

Environmental degradation comes about because societies strive for a better way of life—nicer, more comfortable homes; better clothes; more effective medicines; and more efficient means of transportation, for example. The largest single difference between the "have" and "have-not" countries of the world reflects the abilities of some nations to provide their citizens with better material goods and a larger supply of energy—a higher standard of living—than can other nations.

But the systems of production, distribution, and consumption of material goods and energies always result in the release of waste products into the environment. Polluted air and water, hazardous waste dumps, and dangerous occupations are all examples of the price a society pays for improving its standard of living. And the more the standard of living is improved, the greater the amount of waste released into the environment.

In a perfect world, the benefits and burdens of a high standard of living would be distributed equally throughout a society. Everyone would have equal access to new products and would have to live with increased levels of environmental degradation. But no perfect society exists. Instead, in every form of government devised—socialism, communism, fascism, or democracy—there are haves and have-nots. Some people receive a larger share of the benefits, and others receive a larger share of the burdens. One of the goals of almost any social movement is to obtain a more fair distribution of these benefits and burdens.

How is the distribution of benefits and burdens determined in the real world? One factor is the way in which various groups within the society are viewed. In the United States (as in many other nations), white people tend to have the advantage over people of color, men over women, the physically well over the physically handicapped, heterosexuals over gay men and lesbians, and so on. This pattern is reflected in the wage gap among various groups in the United States. In 2006, for example, the last year for which data are available, white women earned about three-quarters (73.5 percent) as much as men in the United States; black men earned 72.1 percent as much; black women, 63.6 percent as much; Hispanic men, 57.5 percent as much; and Hispanic women, 51.7 percent as much (Infoplease 2008).

The unequal distribution of power and prestige is also reflected in the relative numbers of men and women, whites and nonwhites, and other demographic groups in positions of power. In the 110th Congress of the United States (2007–2009), for example, there are 16 women (16 percent; all white) in the Senate compared to 84 men (84 percent), whereas the percentage of women in the general population is about 52 percent. At the same time, there are 74 women (17 percent) in the House, of whom 15 (3 percent) are African American, 3 (0.6 percent) are Asian Pacific Americans, and 7 (1.6 percent) are Hispanic Americans (Amer 2008).

Under these circumstances, it is not difficult to see why white males tend to receive the highest proportion of material goods available in the U.S. society. And the burdens—such as increased amounts of hazardous wastes, polluted air and water, and greater occupational hazards—that develop as a result of an improved lifestyle for the more privileged are likely to fall on poor people of color. The environmental justice movement has fought to make that fact abundantly and unquestionably clear.

NIMBY, PIBBY, and LULUs

Beyond political and social philosophy, however, there are some concrete reasons that the burdens of environmental degradation have fallen disproportionately on communities of color and on people of low economic status. In the first place, decisions as to where polluting industries, hazardous waste dumps, radioactive storage sites, and other environmentally undesirable sites are to be located are commonly made by governmental bodies, such as city councils, planning departments, or zoning committees. It is not uncommon for such bodies to consist of a disproportionate number of the local "power structure," which usually means white males. The tendency in many cases may be, then, for environmentally undesirable sites to be located in "their" neighborhoods rather than in "our" neighborhoods.

This tendency is reinforced by the reluctance of ordinary citizens to have such sites located in their neighborhoods. One of the most familiar phrases in land-use issues in the last few decades has become "not in my backyard," or NIMBY. The phrase reflects the fact that most people recognize that undesirable factories, prisons, halfway houses, hazardous waste sites, and other "locally unwanted land uses" (LULUs) are now a reality in our society. Given our way of life, they have to exist. But few people want them near their homes. They would prefer to have them constructed "somewhere else." In many cases, this outlook has led to a new attitude, the "put it in the blacks' backyard," or PIBBY, syndrome.

Concerns about Environmental Issues among People of Color

Consonant with the NIMBY principle has been the belief, spoken or unspoken, that African Americans and other minorities are not concerned about environmental issues. Some people have argued that people of color have too many other concerns—jobs, housing, and health problems, for example—to care about the surroundings in which they live. A number of research studies have been conducted to determine the factual accuracy of this attitude.

In his study of environmental problems in five largely African American communities, Robert Bullard found that residents were very much interested in and concerned about such issues. Overall, more than half of the respondents in Bullard's survey had taken part in some form of environmental activism, such as writing a letter to or telephoning an official about an issue, signing or circulating a petition, attending or organizing a meeting in one's home, marching in a demonstration, attending a public meeting, or helping to raise funds (Bullard 2000; also see Buttel 1987; Lowe and Pinhey 1982; Taylor 1989).

An associated finding of the Bullard study was that environmental activism among African Americans tended to take place not within traditional environmental groups but through other organizations, such as churches, social clubs, labor unions, civil rights groups, or parental groups. On average, 16.3 percent of Bullard's subjects belonged to an environmental group defined as such, whereas memberships were 76.5 percent in churches, 27.9 percent in community involvement groups, and 23.1 percent in parent groups (Bullard 2000).

More recent studies confirm the evidence adduced from this earlier research. In 2003, Paul Mohai reported the results of his review of data from three sources: the University of Michigan National Opinion Research Center Detroit Studies from 1990 and 2002, the findings of Urban League studies collected over a 20-year period, and the *People of Color Environmental Groups Directory*. Mohai's analysis covered not only environmental attitudes of African Americans but also factors such as lifestyle choices, political actions, environmental group memberships, and the environmental voting records of African American legislators. He found four major trends:

- In most respects, African Americans are more likely to make lifestyle choices that are better for the environment than do whites: They consume less meat, buy more pesticide-free foods, and drive less. In a few categories, they trail whites. For example, African Americans tend to recycle less than do whites.
- No statistical difference exists between the number of African Americans and the number of whites who join environmental groups. The major difference between the two groups is that the former tend to affiliate with local grassroots organizations while the latter join more traditional, "mainstream" environmental groups.
- African Americans are more likely to express concerns about environmental issues that affect them directly, such as exposure to lead, high noise levels, and abandoned or run-down properties. Mohai points out that these differences probably reflect the environment in which African Americans live compared to white neighborhoods.
- African American legislators are significantly more likely to support environmental legislation than are their white counterparts. The rate for African American legislators ranges from about 75 to 85 percent; for white Democrats, from about 60 to 80 percent; and for white Republicans, from about 20 to 40 percent.

Mohai concluded that "environmental issues are not 'luxury' issues to African Americans. Survey results such as these demonstrate that environmental quality issues are a priority on many different levels" (2003, 25).

Two later studies confirmed Mohai's conclusions. In 2005, Matthew Whittaker of the University of Iowa, Gary M. Segura at the University of Washington, and Shaun Bowler at the University of California at Riverside reported on their study of Latinos in California over a period covering just over two decades. They found that the subject population evidenced greater interest in environmental issues than did whites, but only on issues that related directly to their own lives (Whittaker, Segura, and Bowler 2005). A year later, Robert Emmet Jones of the University of Tennessee and Shirley A. Rainey of Austin Peay State University described their study of 247 people living in a "highly polluted" community of color. They found that their respondents had a very different perception of which environmental issues were of concern to them than did a control group of whites and that they showed a greater level of interest and participation than did whites (Jones and Rainey 2006).

Access to Tools of Protest

The most important single reason that LULUs end up in neighborhoods with a majority of people of color or of people in lower economic strata may be the difference in access that various groups of people in the United States have to the political process. One might argue that a community of color has the same opportunity to protest the siting of a hazardous waste site or a polluting industry as does any other community. It can take its case before a zoning board or bring suit in a court of law to prevent the siting of an undesirable facility, for example.

The reality is, however, that people of color and poor people probably lack the experience, the training, or the financial resources to engage in the same kinds of battles that affluent white communities routinely use to keep their neighborhoods free of environmentally offensive sites. Mohai and Bryant have summarized the disadvantages under which communities of color operate when confronted with environmental hazards. In the first place, the residents of such communities are usually unaware of policy decisions that impact them, and they generally do not have the resources, time, money, experience, or knowledge of the political system to know how to react to those policies. In addition, minority communities tend to be underrepresented on bodies that make and administer policies determining environmental quality (Mohai and Bryant 1992).

Job Blackmail

The siting of LULUs in communities of color or low-income communities often presents a difficult dilemma. The construction of a factory or a hazardous waste site may mean new jobs and an expanded tax base for such communities, an important economic incentive when unemployment may be high and municipal services minimal. When Chemical Waste Management, Inc. (CWM), purchased an existing hazardous waste landfill in Sumter County, Alabama, in 1978, for example, it worked hard to convince local residents of the landfill's economic benefit to the community. A chart the company made public, "How Much Will We Get from Chem Waste?" showed 21 different governmental and nongovernmental bodies—for example, the Sumter County General Fund; the Sumter County Board of Education; the Sumter County Water Authority; the Livingstone University General Fund; the towns of Gainesville, Geiger, and Emelle; the cities of Livingstone and York; and the Sumter County Historical and Preservation Society, the Sumter County Library System, and the Sumter County Fine Arts Council—receiving tax benefits from the landfill. Overall, CWM was to become the largest single taxpayer in the county, producing about half of all public funds collected (Bailey and Faupel 1992).

Poor communities find it difficult to reject an economically productive facility such as this one, even if it might present serious environmental and health hazards. Individual workers face a similar dilemma. It seems unlikely that anyone would choose to work in a hazardous waste landfill or a factory in which noxious fumes are constantly released, but a dangerous job may be better than no job at all.

This form of job blackmail is hardly new to minority workers. Throughout history, African Americans, Asian Americans, Hispanic Americans, or other minority groups have been given jobs that whites would not want or take. Migrant workers probably do not choose to work 12-hour days in hot, dry fields where they are exposed to dangerous pesticides, for example, because they enjoy the outdoor life. They take such jobs because they may be the only ones open to them. Probably the earliest comprehensive discussion of job blackmail can be found in a book by Richard Kazis and Richard L. Grossman, *Fear at Work: Job Blackmail, Labor, and the Environment*, in which they first use the term (Kazis and Grossman 1982).

In many cases, the assignment of minorities to dangerous and difficult jobs has been justified by pseudoscientific theories about differences in physical or biological traits among various ethnic groups. For many years, for example, the iron and steel industry assigned African American men to work at its coke ovens because of the supposed ability of African Americans to withstand heat better than whites. Similarly, workers with dark skin were once more frequently assigned to work with caustic chemicals in some industries because job-induced skin irritations would supposedly be less noticeable and less objectionable than they would be with lighter-skinned workers (Davis 1977).

Job blackmail on a somewhat more comprehensive scale is sometimes referred to as "economic blackmail," a posture taken by both industry and government. One form of the argument is that low-income and minority communities should think seriously about objecting to environmental problems caused by a polluting industry. If you make too much trouble, that argument goes, the industry will go somewhere else, and you will all be left without jobs.

One of the widest applications of economic blackmail in the United States historically has been the use of Native American lands for the mining of coal, oil, natural gas, uranium, and other valuable minerals; for the construction of energy-generating plants; and for hazardous waste disposal sites. In such cases, corporations have traditionally made tribes very attractive financial offers to allow them to build facilities that probably could not be built elsewhere or at much greater expense. The consequence of this trade-off has been the accumulation of considerable wealth for some tribes, accompanied by a significantly diminished environment with consequent health risks to individuals and communities.

Although Native American tribes have become much more sophisticated in dealing with the issue of economic blackmail, temptations continue. In 2007, the Burnham Chapter of the Navajo Nation in the Four Corners region of the Southwest (where Arizona, Colorado, New Mexico, and Utah meet) was confronted with the proposed construction of the Desert Rock power plant, a joint project of the tribe's Diné Power Authority and Sithe Global, a private, multinational company based in Houston. Diné and Sithe Global noted that the plant would create 200 new jobs for the community and that an additional 200 jobs would be created at the coal mine supplying the plant its fuel. In addition, the new plant would produce far less pollution than others of its kind in the area.

Although the planned construction seemed a win-win situation for everyone, environmental activists were less certain. They pointed out, first, that a number of opportunities for alternative energy generation facilities were available, such as those involving the use of solar power. In addition, they wondered why the owners of the plant selected one of the most impoverished, Indian-occupied lands in the area on which to build rather than choosing land occupied primarily by middle-class whites. Finally, they suggested that reducing pollution significantly does not solve the environmental health problems already faced by Native Americans in the area. As one observer noted, "just because you go from 10 quarts of whiskey a week to four doesn't mean you're clean and sober," and reducing pollution even by half still results in unacceptable levels of noxious gases being released into the atmosphere (Binkly 2007).

Job blackmail and economic blackmail present, therefore, a damned-if-you-do and damned-if-you-don't choice for both individuals and communities. In his article "Environmental Blackmail in Minority Communities," Robert Bullard pointed out that many polluting facilities such as paper mills, chemical plants, heavy metals operations, and waste disposal and treatment facilities saw minority communities as a logical choice for their facilities. They assumed that such communities reasoned as Third World communities might, believing that development of any kind, no matter how dangerous, is better than no development at all. Foul-smelling air and polluted water were, they reasoned, a fair trade-off for jobs and security (Bullard 1992).

An intriguing point about the job/economic blackmail argument is that it often presents a false dichotomy, forcing individuals and communities to choose between two apparently incompatible alternatives. But those alternatives are not necessarily incompatible. In some cases, industries that have said they could not survive unless their demands were met did, indeed, survive (and often prospered), even though they were not able to get their way with a community. One case sometimes mentioned involves the paper industry in Maine, which claimed that the provisions of the Federal Water Pollution Control Amendments (the so-called Clean Water Act) of 1972 would put it out of business because it would be forced to install new pollution control devices. In fact, there appeared to be little economic impact on the industry once it complied with the legislation, and it was not until a decade later that employment actually dropped—as a result of new automated production methods (Sayen 1995).

Other observers point out that meeting new environmental regulations can even benefit an industry. By switching from older, outmoded systems to newer, more-efficient systems, industries can create new jobs and realize unexpected profits. When speaking about pollution problems in his Cincinnati neighborhood, for example, activist Mike Henson reflected on the fact that a local industry, Queen City Barrels, now brags on its Web site about having the best pollution control equipment available in operation at its facility. But, he points out, the company installed the equipment only because of pressure by community activists and, eventually, new city regulations (Pierce 2007).

The Effects of Environmental Regulations

The picture painted thus far of the environmental inequities faced by people of color and poor people seems bleak. Yet, the United States has developed a much more enlightened view of the environment in the past three decades. People in general are probably much better informed and more concerned about the environment, and federal, state, and local governments have become much more aggressive about passing legislation and enforcing regulations that protect the environment. Do these changes suggest that progress is being made in dealing with the disproportionate exposure faced by people of color and poor people in the United States?

It would be difficult to argue that no overall improvement has been made as a result of such changes. Somewhat remarkably, however, it is also clear that many new environmental regulations have actually increased the environmental hazards to which minorities are exposed. Again, the benefits of environmental regulations for some groups of people have resulted in an increased burden for other groups of people.

First, new programs to protect the environment clearly cost money. Those costs are borne by all taxpayers in the society, poor as well as rich. They are borne by all taxpayers whether or not they benefit directly from the environmental program in question. Programs to protect national wildlife areas or national seashores are paid for by African Americans, Hispanic

Americans, and Native Americans, as well as by whites. Every time an environmental regulation raises the cost of an automobile because of new pollution control devices, increases the cost of housing because of new land-use regulations, or raises the cost of food products because of new pesticide-use rules, people of lower-income groups pay a disproportionate percentage of that cost.

In other words, the economic costs of new environmental regulations are more likely to have a more significant effect on low-income people than on high-income people. If the costs of environmental regulations average out to $500 per person nationwide, those costs will be more detrimental for someone making $10,000 per year than for someone making $100,000 per year.

One attempt to measure the economic impact of the air pollution policy in the United States was conducted by Leonard P. Gianessi, Henry M. Peskin, and Edward Wolff. This study was used to determine the costs and benefits that could be anticipated from the full implementation of the federal Clean Air Act. The researchers found that the effects of the Clean Air Act were regressive; that is, people of low income would pay proportionally more than would those of high income. For those with an income ranging between $3,000 and $8,000 per year, the net costs of the Clean Air Act were about 0.9 to 1.0 percent of their income. In contrast, those with incomes of more than $8,000 uniformly experienced a net cost of about 0.5 percent of their income (Gianessi, Peskin, and Wolff 1979; also see Johnson 1980).

The costs of environmental regulation can also be measured in terms of exposure to regulated pollutants. In his doctoral thesis at the University of California at Berkeley, Michel Gelobter analyzed the amount of air pollution to which individuals in selected urban areas were exposed in the period between 1970 and 1986. He found that all the racial and all the income groups he studied were exposed to lower levels of ambient air pollutants at the end of that period than they were at the beginning. However, Gelobter found differences in exposure among both income and racial groups. For example, in 1970, 1975, 1980, and 1984, the average white resident of the urban areas studied was exposed to 100, 75, 75, and 62 micrograms of total suspended particulates per cubic meter of air, respectively, whereas the comparable figures for nonwhites during those four years were 122, 85, 82, and 70 micrograms of total suspended particulates per cubic meter of air, respectively. In an analysis of the "relative benefit from air quality improvements" that included a variety of measures, Gelobter found that whites and nonwhites rated almost exactly the same in 1970. Over the next 15 years, however, the difference in relative benefits diverged until whites were approximately 5 percent "better off" according to these measures. Gelobter later summarized the results of his research in a chapter for the book *Race and the Incidence of Environmental Hazards: A Time for Discourse*, edited by Bunyan Bryant and Paul Mohai (Gelobter 1992).

A number of studies have produced another interesting finding: the U.S. Environmental Protection Agency (EPA), the U.S. government agency primarily responsible for the prevention and cleanup of polluted sites, has itself been guilty of discriminatory policies and practices in dealing with communities of color and low-economic status. In a study conducted for Clean Sites, Inc., in 1990, for example, Kate Probst found that the EPA was less likely to include rural poor communities on its list of potential Superfund sites than other communities (Probst 1990).

One of the most telling studies on this issue was conducted by Lavelle and Coyle in 1992. These researchers found, first of all, that the EPA was likely to make minority communities wait even longer than low-income communities to receive Superfund listing. The pattern was for low-income areas to wait 11 percent longer than high-income areas, and for minority areas to wait 20 percent longer than white areas. Lavelle and Coyle also found that, after a minority community did receive a Superfund listing, the agency was more likely to institute a program of waste containment than to arrange for toxicity reduction or removal, as was the case with white communities. Finally, for instances of pollution in minority communities the fines assessed by the EPA averaged only 20 percent of those assessed in white communities (Lavelle and Coyle 1993).

Critics have argued that an indication of the EPA's attitudes toward racial issues can be found in the agency's hiring policies. One way in which the agency could demonstrate a greater sensitivity to issues of environmental inequities, they suggest, would be to make greater efforts to include members of minorities on its staff. Yet, according to an agency report published in 1992, less than 10 percent (33 out of 412) of the new management positions filled during the previous year were given to members of a minority. In the following year (1992), 42 of 354 management hires went to minorities. In comparison, the number of white females hired was 142 (34 percent of new hires) and 126 (36 percent) in 1991 and 1992, respectively (U.S. Environmental Protection Agency 1992).

In a critique of the nation's environmental protection policies, Robert Bullard has written that these policies have "(1) institutionalized unequal enforcement; (2) traded human health for profit; (3) placed the burden of proof on the 'victims' and not on the polluting industry; (4) legitimated human exposure to harmful chemicals, pesticides, and hazardous substances; (5) promoted risky technologies, such as incinerators; (6) exploited the vulnerability of economically and politically disenfranchised communities; (7) subsidized ecological destruction; (8) created an industry around risk assessment; (9) delayed cleanup actions; and (10) failed to develop pollution prevention as the overarching and dominant strategy" (Bullard 2000, 115).

Environmental Inequities: By Chance or By Choice?

Even if everything said above were entirely true, one critical question remains—the question of intent. Do environmental inequities exist because corporations, governmental bodies, and/or individuals plan for them, or do they come about simply as the result of business and/or technical decisions based on other factors? This question has been at the core of environmental justice issues because, from the earliest days of the movement, the assumption has been that environmental inequities are a manifestation of racism and, perhaps to a lesser degree, class issues. Recall that Benjamin Chavis defined environmental racism as "the deliberate targeting of people of color communities for toxic waste facilities and the official sanctioning of a life threatening presence of poisons and pollutants in people of color communities" (U.S. Congress 1993, 6).

Some people, however, while acknowledging the existence of environmental inequities, argue that racism is a minor or nonexistent factor. Such inequities come about not because business, industry, and government intend to expose poor and minority communities to environmental insults; instead, these problems arise as a by-product—and an unfortunate by-product that needs to be addressed—of business decisions that have no racial or class component.

This issue is, of course, a crucial one within the environmental justice arena. Perhaps corporations and governments routinely make decisions about the siting of LULUs based strictly on economic factors, such as labor costs, availability of land, and access to transportation. Racial, ethnic, or other social factors might then have little or no impact on their decisions. In such a case, minority communities could expect to work with corporations and governments to find ways of siting such facilities with less disproportionate impact on themselves. The affected communities and the agencies responsible for environmental inequities could work together with trust and confidence in each other.

On the other hand, it might be that factors other than economic matters enter into the siting of LULUs. One could imagine that those responsible for making such decisions hold African Americans, Hispanic Americans, Asian Americans, Native Americans, and poor people in less regard than they do middle- and upper-class white citizens. In such a case, solving the problem of environmental inequities poses a different dilemma. Minority communities would be faced with the task of changing attitudes among those in corporations or government whose respect they currently do not have. Or they might have to use a more confrontational approach to problem solving (such as court cases) to deal with environmental inequities. As a matter of fact, it is the latter position that most of those active in the environmental justice movement appear to hold currently and to have held since the earliest days of the movement.

Still, many observers continue to deny the presence of social attitudes, such as racism, in the origin of environmental inequities. An example is the comment made by Kent Jeffreys, director of Environmental Studies for the Competitive Enterprise Institute, in testimony before the Subcommittee on Civil and Constitutional Rights of the U.S. House of Representatives on March 3, 1993. Jeffreys said that "racism exists. Environmental problems exist. These facts, however, do not reveal whether or not environmental racism is occurring. Regardless of whether any particular case fits the definition of environmental racism, the fact remains that environmental problems—from a minority perspective—are rather trivial in comparison to the larger economic and civil liberty issues: solve these and you have solved most, if not all, of the environmental inequities" (U.S. Congress 1993, 64).

This viewpoint was also reflected during hearings before the Louisiana Advisory Committee to the U.S. Commission on Civil Rights in February 1992. The hearings were held to determine the status of environmental problems in certain parts of the state. At the conclusion of its fact-finding sessions, the committee concluded that "many black communities located along the industrial corridor between Baton Rouge and New Orleans are disproportionately impacted by the present State and local government system for permitting and expansion of hazardous waste and chemical facilities" (Louisiana Advisory Committee 1993, iii). The committee did not specifically attribute the environmental inequities it found to racism, but its final report contains a number of allusions that suggest that the majority of committee members may have held this viewpoint. At one point the report says, for example, that "some black citizens and organizations view with skepticism and distrust some agencies in State government, including the Department of Environmental Quality. These views are held because of poor access to government and an ongoing perception that State government discriminated against them to promote and sustain the interests of industry and business" (Louisiana Advisory Committee 1993, iv).

Perhaps the best view of the committee's mind-set on this issue comes from one member who wrote a dissenting statement to the report. John S. Baker Jr., professor of law at the Louisiana State University Law Center, concluded that "absent from the report is the one finding most clearly supported by the evidence: **Environmental Racism has not been shown to exist in Louisiana**. . . . None of the extensive findings contains anything about, nor could they support a finding of, 'deliberate targeting' or 'official sanctioning'" (Louisiana Advisory Committee 1993, 69 [emphasis in original]). Baker argued that members of the committee knew in advance of the hearings that discriminatory motive would not be found and that the sessions were conducted with other purposes in mind than the search for such a motive. Baker continued in his dissent:

> The overall handling of this project reminds me of the situation that occasionally occurs during a criminal case. Sometimes during voir dire when a potential juror is asked whether he or she can presume the innocence of the defendant and give him a fair trial, the juror will answer quite innocently in terms that effectively say:

"Of course, I'll give him a fair trial before convicting him." Like the prospective juror, the staff and committee members, I believe, are acting with the best of intentions. Nonetheless, the report, and the process that produced it, have blindly stepped over and around the evidence, which so clearly establishes no racially motivated discrimination on environmental decisions in Louisiana. (Louisiana Advisory Committee 1993, 71–72)

The issue being raised in this debate might be characterized as a chicken-and-egg debate: Which came first—the hazardous waste facility or low-income and minority communities? In other words, do industries tend to choose an existing minority and/or low-income community in which to locate because they have less regard for the health and welfare of people living there, or does an industry build a landfill (or other LULU) and low-income and minority people then choose to move into that area?

Some research has been done on this question, but the findings are not yet conclusive. For example, Vicki Been, associate professor of law at New York University School of Law, has reanalyzed the demographics of waste disposal sites previously studied by the U.S. General Accountability Office (GAO) and by Robert Bullard in Houston. Her intent was to pay greater attention to the economic and racial status of an area before a toxic waste landfill was constructed.

Been obtained somewhat conflicting results from her examination of the GAO and Bullard studies. In the former case, she found that the four communities where waste disposal sites were sited were originally both predominantly low-income and minority communities. She also found that neither the proportion of low-income nor the proportion of minority individuals increased after the siting. In the Houston (Bullard) case, Been found that waste disposal sites were occupied predominantly by minority populations, but not by low-income populations, before facilities were installed. After the facilities were opened, however, the percentage of African Americans and of low-income people increased. Been concludes from her research that "research examining the socioeconomic characteristics of host neighborhoods at the time they were selected, then tracing changes in those characteristics following the siting, would go a long way toward answering the question of which came first—the LULU or its minority or poor neighbors. Until that research is complete, proposed 'solutions' to the problem of disproportionate siting run a substantial risk of missing the mark" (Been 1994, 124).

One of the difficulties in conducting studies of this type is the difference between older hazardous waste sites, many now closed, and new or proposed sites. At one time, it was relatively easy to build a hazardous waste site, and older sites can be found throughout the United States. Now, obtaining permission to construct a hazardous waste site is much more difficult. It follows that the factors involved in making the decision as to where to build a newer site might, therefore, be somewhat different now than in the past. One study bearing on this issue was completed by James T. Hamilton. Hamilton found that counties planning to expand waste disposal sites apparently made no distinction among areas on the basis of racial or economic conditions. He also found, however, that sites scheduled for closure were more likely to be located in white areas than in minority communities (Hamilton 1993).

In his summary of the new research needed in the field of environmental justice, Benjamin Goldman returns to the question of causality in the siting of LULUs. He points out that the majority of studies he reviewed make use of correlation analysis, which shows how two variables are mathematically related to each other, but do not demonstrate whether or how one variable is dependent on the other. Thus, the fact that the siting of hazardous facilities is mathematically related to a high density of nonwhite and/or poor populations does not prove that one of these factors causes or is caused by the other (Goldman 1993).

Yet, as Goldman also points out, studies like those by Been and Hamilton deal with the very core of the environmental justice movement. He goes on to say that the basic questions for those concerned about environmental justice are about how the decisions made by legal, social, economic, and other cultural institutions result in disproportionate environmental impacts and what legal remedies are available for resolving those impacts, whether or not one can prove intent by an offending industry.

The Legal Question of "Intent" in Environmental Inequities

Goldman's summary refers to another key issue in the dispute over environmental inequities: the question of intent. Many involved in the environmental justice movement would like to demonstrate that environmental harm has come about as a result of decisions and actions taken by business, industry, or governmental bodies and then to obtain some relief from that harm from the appropriate body. That is, an environmental justice organization might like to prove that Company X placed its hazardous waste landfill adjacent to a predominantly African American community because that was the easiest decision for the company to make. The organization might then argue that health effects to the community resulting from the company's decision were the company's responsibility, and that the company should be forced to pay a financial penalty for its decision.

And in some cases, this approach has worked, and companies have paid large fines or made large financial settlements to members of communities harmed by their activities. But such instances are relatively rare. One reason is that a stringent standard of proof is often required by courts in cases of environmental harm. This standard was established in the landmark decision in *Washington, Mayor of Washington, D.C., et al. v. Davis et al.* In that case, the U.S. Supreme Court ruled that a plaintiff must be able to prove that harmful actions taken by an individual, a group, or a corporation are intended to cause harm to the plaintiff and not that the harm occurred as an unexpected by-product of the action. This legal standard is a difficult one to employ since it requires that a plaintiff somehow find out what was in the minds of a businessperson or government official and then demonstrate "intent" to the court.

The hurdles created by the intent standard have meant that many environmental justice activists have used means other than legal challenges to deal with environmental issues. As one example, the Lumbee Indians of Robeson County, North Carolina, were involved in an extended dispute with the GSX Corporation, which planned to build a hazardous waste facility on lands sacred to the Lumbee. The tactics used by the Lumbee were not legal challenges but a more personal campaign featuring traditional Native American dance and music at public hearings, leafleting at churches, and participation in all planning sessions held on the GSX request (Austin and Schill 1994). Grassroots activities such as these have been common and successful in situations where long, expensive, and complex legal challenges would have held much less promise.

Responding to Environmental Inequities

Some obvious methods for dealing with environmental inequities have been suggested thus far. Some communities have found that direct political action—protests and demonstrations, for example—have accomplished their goals. Others have sought relief in the courts. A debate remains, however, as to the best mechanism by which a community dealing with a LULU can protect itself against the dangers of such a facility.

On the one hand, some people believe that environmental justice is, to a large extent, a problem with which governments have to deal. New and better laws and stronger enforcement of existing laws can prevent future environmental inequities and can help deal with instances now existing. Robert Bullard, for example, has called for government to take five steps to ensure environmental justice: creating national legislation modeled on civil rights acts, eliminating existing environmental hazards, shifting the burden of proof for guilt or innocence in environmental inequity cases from affected communities to polluting industries, modifying the current legal standard of intent in dealing with cases of environmental inequities, and providing compensation and assistance to communities most seriously affected by environmental inequities (Bullard 1994).

An opposing view to the solution of environmental inequities calls for greater dependence on market forces in the private sector and less on government involvement. Companies that benefit from the use of land for the construction of LULUs should be required to pay a fee for this privilege. That fee would then be transferred to minority and poor communities on whom the burdens of the LULU siting fall.

An example of this approach is legislation proposed in the state of Wisconsin in 1981. This legislation required negotiations between a company wishing to build a hazardous waste facility and the municipality in which the facility would be located. Three advantages of the Wisconsin plan, as noted by Christopher Boerner and Thomas Lambert of the Center for the Study of American Business at Washington University in St. Louis, are that the statute "clearly specifies 'the players of the game'—that is, who negotiates with whom. Both developers and potential host communities are required to establish negotiating committees, with the rules regarding these representatives explicitly set forth in the statute. Secondly, the legislation assures that these 'players' will not only negotiate, but also that the results of their negotiations will be legally binding. . . . Finally, the Wisconsin statute provides a 'back-up plan'—a way to arbitrate siting decisions should negotiations fail or should one party refuse to cooperate" (Boerner and Lambert 1995, 99).

The Current Status of the Environmental Justice Movement in the United States

The March 2007 United Church of Christ report *Toxic Wastes and Race at Twenty* includes a timeline of important events in the environmental justice movement from 1987 to 2007 (Bullard et al. 2007, chap. 2). That timeline provides a succinct review of the progress (or lack of progress) in dealing with environmental inequities in the United States and a hint as to the problems and challenges still facing the movement. A few of the generalizations that can be drawn from the timeline are described in the following sections.

Environmental Inequities Exist

Despite all the debate that has taken place over whether environmental inequities exist, what groups of people are most affected by such inequities, whether inequities occur as the by-product of industrial development or because of corporate intent, and other details, the fact remains that very few experts any longer doubt that low-income and minority communities in the United States suffer disproportionate health risks because of their exposure to hazardous waste dumps, nuclear processing plants and disposal sites, unsafe occupational settings, petrochemical and chemical industries, and other facilities that emit harmful products into the environment. Report after report begins with some mention and acknowledgment of this reality. The 2003 report by the U.S. Commission on Civil Rights, *Not in My Backyard: Executive Order 12898 and Title VI as Tools for Achieving Environmental Justice*, for example, notes that "clearly, race and class play significant roles in environmental decision-making; moreover, communities of color and low-income communities are disproportionately affected by siting decisions and the permitting of facilities. Siting and permitting decisions are not, however, the sole sources of environmental concerns in these communities. Exposure to lead-based paint, diesel emissions, noise, odor, and other pollutants also diminishes the health of these communities." Minority and low-income communities are exposed to multiple pollutants from multiple sources, the report continued; however, there has been insufficient data collection and scientific research identifying the health risks created by these multiple exposures (U.S. Commission on Civil Rights 2003, 27).

Two years later, the U.S. Government Accountability Office (GAO) issued a report on clean air rulemaking by the EPA. Its letter of transmittal about the report began with the statement that "[l]ow-income and minority populations are disproportionately exposed to air pollution and other environmental risks, according to Environmental Protection Agency (EPA) studies" (U.S. Government Accountability Office 2005, 1).

Shortly after the GAO report appeared, the Associated Press reported on "a little-known government research project" dealing with the prevalence of environmental inequities using data from the 2000 census. That report indicated that 79 percent of African Americans were more likely to live in neighborhoods where pollution could pose health risks than are whites. It concluded that "residents in neighborhoods with the highest pollution scores also tend to be poorer, less educated and more often unemployed than those elsewhere in the country." The study also found that similar patterns do not hold nationwide for Asians and Hispanics, although they do occur in certain states (12 states for Hispanics and 7 for Asians) (Pace 2005).

Federal, State, Regional, and Local Governmental Agencies Have a Responsibility to Find, Study, and Deal with Cases of Environmental Inequities

The first major governmental action dealing with environmental inequities was Executive Order 12898, signed by President Bill Clinton on February 11, 1994. (Title VI of the Civil Rights Act of 1964 has also become an important tool in dealing with environmental justice issues, although it was not adopted specifically for that purpose.) Since the time of Clinton's proclamation, a number of other executive, legislative, and regulatory orders have been adopted for dealing with problems of environmental injustices. In 2003, for example, the New York State Department of Environmental Conservation adopted a new policy requiring the inclusion of environmental justice considerations in the issuance of new permits. The new policy amended the existing permitting process by "identifying potential environmental justice areas; providing information on environmental justice to applicants with proposed projects in those communities; enhancing public participation requirements for proposed projects in those communities; establishing requirements for projects in potential environmental justice areas with the potential for at least one significant adverse environmental impact; and providing alternative dispute resolution opportunities to allow communities and project sponsors to resolve issues of concern to the community."

A year later, New Jersey governor James McGreevey became one of the first state executives to take action on environmental justice issues. On January 19, 2004, he issued an executive order similar to Clinton's order of 1994 that required all state departments, agencies, boards, commissions, and other bodies to "provide meaningful opportunities for involvement to all people regardless of race, color, ethnicity, religion, income, or education level" ("Environmental Justice Executive Order [New Jersey]" 2004). The order included a number of specific provisions, such as requiring that notices on health and environmental issues be made available in Spanish as well as in English, studying existing and proposed industrial and commercial facilities located near minority and low-income communities to determine their impact on the health of individuals in those communities, developing advisories concerning consumption of fish that may contain high levels of hazardous materials, creating a strategy for dealing with health problems faced by low-income and minority communities as a result of exposure to particulate matter in the air, and creating or reconstituting a number of state agencies with responsibility for environmental justice issues.

In 2004, the section of Individual Rights and Responsibilities of the American Bar Association and the Public Law Research Institute of the University of California's Hastings College of Law issued a comprehensive report on the actions taken in all 50 states in the field of environmental justice. The report shows that a host of tools has been developed by the states for dealing with problems of environmental inequities, ranging from statutes and executive orders to policy statements and special training programs. As one might expect, dramatic differences exist in the extent to which states (if any) acknowledge the existence of environmental inequities and/or take action to deal with those inequities, as well as in the tools they have developed to use in environmental justice problems. The examples that follow demonstrate the differences (all information as of 2004 and obtained from Bonorris 2004).

Alaska has no laws or regulations dealing with environmental inequities.

Arizona has no state laws on environmental justice, but the state's Department of Environmental Quality has expressed a commitment to monitor and act on issues of environmental inequities that come to its attention. It maintains one full-time employee to deal with such issues.

California has one of the most extensive and comprehensive collections of laws and regulations for dealing with environmental justice issues of any state. The first such law was passed in 1999 designating the Governor's Office of Planning and Research as the lead agency for environmental justice programs and directing the California Environmental Protection Agency to take a number of actions to ensure that the state is notified of and takes actions on specific instances of environmental injustice that occur in the state. During the three-year period from 2001 to 2004, the state legislature passed eight more bills dealing specifically with environmental justice concerns, such as diesel engine emission reduction programs, cleanup programs for small parcels of land contaminated with hazardous wastes, models for the siting and operation of hazardous waste disposal sites, and a program of small grants for communities dealing with environmental and health problems specific to their neighborhoods.

Florida has, like most other states, adopted laws and regulations ensuring that problems of environmental inequities are discovered and acted upon. But it has also developed a somewhat unusual model in that most of the actions taken to deal with environmental justice issues take place outside of state agencies and within the state's university system. Much of this work is carried out through five university divisions: the Environmental Sciences Institute, the Center for Environmental Equity and Justice, the College of Pharmacy and Pharmaceutical Sciences, the Institute of Public Health at the Florida Agricultural and Mechanical University, and the Interdisciplinary Center for Brownfield Rehabilitation Assistance at the University of South Florida. These divisions are responsible for conducting research and working with communities on problems of environmental injustice.

Indiana stands out among states as having been especially proactive in dealing with its problems of environmental inequities. The state's Department of Environmental Management (IDEM) adopted a strategic plan for environmental justice in August 2001. Key elements in that plan include identifying geographic areas in which environmental inequities of significance exist and developing procedures by which communities can make their concerns known to state agencies. Although the state's approach to environmental justice has been recognized as a "model for change" by the National Academy of Public Administration (NAPA), some internal problems remain. Since other state agencies responsible for air and water quality and solid waste management continue to function independently of IDEM, the latter agency's ability to bring about significant change is somewhat limited (see National Academy of Public Administration 2002, 37–43; Bonorris 2004, 24).

Kentucky is an example of a state with minimal environmental justice provisions. A 2002 law requires that the Kentucky Regional Integrated Waste Treatment and Disposal Facility Siting Board consider the social and economic impacts on communities and the environment of any proposed new or existing waste treatment or disposal site. The state has virtually no other statutory or regulatory provisions for dealing with environmental inequities.

New Jersey was another state recognized by the 2002 NAPA study of "models of change" in environmental justice. The NAPA report especially applauded the state's decision to include environmental justice considerations in its broader plan to deal with an array of state problems, including urban sprawl, availability of open space, preservation of agricultural lands, rehabilitation of brownfield sites, and redevelopment of older, industrialized cities. Well before Governor McGreevey's executive order on environmental justice (see above), Robert C. Shinn Jr., commissioner of the state's Department of Environmental Protection, issued a number of orders dealing with environmental inequities. The first, issued in 1998, created the Environmental Equity Task Force to develop policy and make recommendations for environmental justice in the state. The second, issued in 1999, was designed to carry out the task force's recommendations. And the third, issued in 2000, set out an extensive and comprehensive state environmental justice policy that reflected the recommendations of the task force.

Virginia first recognized the problem of environmental inequities in 1993 when the legislature adopted a joint resolution requiring the Joint Legislative Audit Review Commission (JLARC) to study the "siting, monitoring, and cleanup of solid and hazardous waste facilities, with an emphasis on how waste facilities affect minority communities" (Bonorris 2004, 55).

The Federal Government Has Not Been a Strong Advocate for Environmental Justice

Early in the environmental justice movement, many people were confident that they possessed two powerful tools to challenge cases of environmental inequities: President Clinton's Executive Order 12898 and Title VI of the Civil Rights Act of 1964. They believed the executive order could be used by federal officials (and, perhaps, by example, state, regional, and local officials) to use their regulatory powers to find and fight cases in which low-income and minority communities are exposed to disproportionate health risks from hazardous waste dumps, occupational hazards, nuclear mining operations, and other hazardous facilities. The executive order created a new federal framework for dealing with environmental justice and required all federal departments to develop strategies for including the principles of environmental justice within their overall philosophies and day-to-day operations.

From the outset, however, that expectation met with mixed results. Some departments appointed task forces to study the ways in which the executive order could be implemented within their own agencies, appointed men and women with specific responsibilities for environmental justice activities, offered training classes in dealing with environmental justice issues, offered grants for research and action programs on environmental justice, and otherwise complied with both the spirit and the letter of the executive order. Other departments moved more slowly or, in some cases, hardly at all.

The potential for action using the executive order was reduced significantly with the accession of George W. Bush to the presidency. Many observers believed that Bush was not particularly interested in environmental justice, and his appointees to the EPA tended to reflect that position. Over the years, the first efforts to include environmental justice concerns within departments faded or disappeared entirely. The Interagency Working Group established by the executive order became moribund and, by 2008, touted a conference on environmental justice as its main accomplishment and indication that it was still active and moving aggressively on environmental justice issues (Gogal, 2008).

During the first decade of the 21st century, a number of governmental and nongovernmental organizations and agencies began to issue reports highlighting the failure of the federal government to take environmental justice concerns seriously, often criticizing specific agencies for their tendency to ignore precise instructions in the executive order. For example, the EPA's own Office of Inspector General (OIG) issued a report in March 2004 reviewing the agency's efforts to implement Executive Order 12898. The OIG concluded that the "EPA has not fully implemented Executive Order 12898 nor consistently integrated environmental justice into its day-to-day operations. EPA has not identified minority and low-income, nor identified populations addressed in the Executive Order, and has neither defined nor developed criteria for determining disproportionately impacted" (Office of Inspector General 2004, i).

The OIG report goes on to suggest that, although the agency has been actively involved in implementing Executive Order 12898 for 10 years, it has not developed a clear vision or a comprehensive strategic plan and has not established values, goals, expectations, and performance measurements. The agency did make an attempt to issue an environmental justice toolkit, endorsed environmental justice training, and required that all regional and programmatic offices submit "action plans" to develop some accountability for environmental justice integration.

In the absence of environmental justice definitions, criteria, or standards from the agency, the OIG report concluded, many regional and program offices have taken steps, individually, to implement environmental justice policies. This has resulted in inconsistent approaches by the regional offices. Thus, the implementation of environmental justice actions is dependent not only on minority and income status but on the EPA region in which the person resides. A comparison of how environmental justice protocols used by three different regions would apply to the same city showed a wide disparity (Office of Inspector General 2004, i–ii).

Two years later, the OIG issued a second report on the EPA's internal implementation of the provisions of the executive order and found that very little had been accomplished (Office of Inspector General 2006). Reviews from outside the EPA found similar results. In July 2005, a report by the GAO found that the EPA had essentially ignored Executive Order 12898 and any consideration of environmental justice concerns when it issued three new regulations on air quality.

Perhaps the most striking feature of the Bush administration's approach to environmental justice has been its attempt to redefine the meaning of the term in such a way as to subvert its fundamental intent, namely, to focus on the disproportionate environmental inequities faced by minority and low-income communities. It decided to define environmental justice as "the fair treatment and meaningful involvement of all people regardless of race, color, national origin, or income, with respect to development, implementation, and enforcement of environmental laws, regulations, and policies" (U.S. Environmental Protection Agency, 2008). In a June 2002 memo to employees of the EPA, the agency's Office of Environmental Justice offered a redefinition of its mission. "Senior management should recognize," the memo said, "that the environmental justice program is not an affirmative action program or a set-aside program designed specifically to address the concerns of minority communities and/or low-income communities. To the contrary, environmental justice belongs to all Americans and it is the responsibility of Agency officials, as public servants, to serve all members of the public" (as quoted in Office of Inspector General 2004, 10). In other words, the purpose of the EPA's environmental justice program was not to provide relief to those low-income and minority communities who suffer disproportionate health risks from exposure to hazardous materials but to distribute the risks and the benefits of environmental justice (whatever the term now means) to all Americans equally.

The OIG rejected this action by the EPA's Office of Environmental Justice. "We believe the Agency is bound by the requirements of Executive Order 12898," the OIG report said, "and does not have the authority to reinterpret the order. The Acting Deputy Administrator needs to reaffirm that the Executive Order 12898 applies specifically to minority and low-income populations that are disproportionately impacted" (Office of Inspector General 2004, ii). The EPA's efforts to redefine environmental justice also produced a firestorm of objection from environmental justice groups and from legislators. In August 2005, for example, more than 70 Democratic members of Congress wrote the EPA objecting to its attempts to redefine the meaning of environmental justice and its failure to carry out the objectives of Executive Order 12898 (Featherstone 2005).

Complaints Filed with the EPA Have Not Been an Effective Means of Dealing with Environmental Inequities

During the early 1990s, activists decided that the best way to challenge environmental inequities faced by low-income and minority communities was to file complaints with the EPA based on the equal rights provisions contained in Title VI of the Civil Rights Act of 1964. In the decade following 1993, when the first such complaint was filed, the EPA received 143 complaints. Of that number, the EPA rejected 82 cases outright, primarily because they did not meet regulatory requirements or because the complaints were made against companies or agencies that did not receive EPA funding (a requirement for such claims). As of November 2003 (the last date for which complete data are available), 29 complaints were still pending before the EPA. Of the 32 cases that have been closed, 2 were settled informally, and 2 were referred to other agencies. Ten more cases became moot because the complaint was dropped, the challenged permit was withdrawn or denied, or some other mitigating action was taken. Of the 18 cases on which the EPA took action, it rejected the complaint in every case. In most cases, the EPA found that complainants did not show a direct causation between hazardous wastes or emissions and human health problems, did not demonstrate that low-income or minority communities had been disproportionately affected by exposure to hazardous wastes, or dealt with nonessential issues (such as not providing information in languages other than English). As one expert in the field has concluded, "it now appears that the EPA's complaint procedure is not providing an effective source of redress . . . unless some of the 29 cases still pending have different outcomes than the 105 already closed" (Gerrard 2003, 6; also see Toffolon-Weiss and Roberts 2004, 265; Pace 2005).

Environmental Justice Advocates Have Largely Failed to Obtain Redress from Inequities through the Judicial System

Another avenue activists can use to pursue environmental justice complaints is through the courts. Early in the movement's history, some authorities believed that court cases could be brought based on the equal protection clause of the 14th Amendment to the U.S. Constitution. That amendment says that "no state shall . . . deny to any person within its jurisdiction the equal protection of the laws." Since disproportionate exposure to hazardous materials would seem to violate this clause, a number of complaints were filed in courts claiming discrimination. These complaints, however, failed almost universally (for details of this history, see Peter 2001).

Eventually, attorneys decided that environmental justice clients were more likely to achieve success suing on the basis of Title VI of the Civil Rights Act of 1964. That is, they planned to sue corporations or governmental agencies on the basis that their clients' civil rights were being violated by disproportionate exposure to hazardous and toxic materials. A number of court cases were filed under this rubric, but without much success. An example of such cases was *Sierra Club v. Gates*, a suit filed in the U.S. District Court for the Southern District of Indiana in early 2007. A group of environmental organizations sought to enjoin the U.S. Department of Defense and its private agent, Veolia Environmental Services, from incinerating 1.8 million gallons of the chemical warfare agent VX, which is converted into the chemical known as hydrolysate before disposal. Veolia planned to conduct this operation in Port Arthur, Texas, on land adjacent to a largely low-income, African American community.

That neighborhood already faced a number of environmental hazards from adjoining refineries and chemical facilities and had been found to have an unusually high rate of respiratory problems; ear, nose, and throat disorders; asthma; cancer; and liver and kidney diseases compared to control groups in nearby Galveston. Residents blamed their health problems on exposure to hazardous wastes produced by the nearby plants and feared that the planned incineration of VX would only exacerbate those problems.

On August 3, 2007, Judge Larry J. McKinney denied the request for an injunction against the Department of Defense and Veolia, saying that the army had made its decision to incinerate the VX materials on scientific principles and that the environmental groups stood little chance of success in pursuing their case. The environmental groups then sought the aid of state and federal lawmakers in an effort to prevent the incineration. Those efforts also failed, and as of late 2008, almost all of the VX had been destroyed in Port Arthur.

Perhaps the most promising court ruling for proponents of environmental justice came on April 19, 2001, when U.S. District Court judge Stephen Orofsky ruled for the plaintiffs in *South Camden Citizens in Action v. New Jersey Department of Environmental Protection* (145 F. Supp. 2d 446 [2001]). Judge Orofsky decided that the state environmental agency had failed to consider the potential disproportionate and adverse effects posed by air emissions from the proposed St. Lawrence Cement Company's plant in South Camden. The legal basis of Orofsky's decision rested in draft guidelines that the EPA had prepared for such cases and on the assumption that individuals have the right to sue to enforce regulations that have disparate effects on communities under Title VI.

Environmental justice advocates were delighted with the ruling, one of the first upholding the general philosophy under which Title VI could be used to deal with environmental inequities. That delight lasted only five days, however. On April 24, The U.S. Supreme Court issued its decision in the case of *Alexander v. Sandoval*, a case unrelated to environmental justice substantively (it dealt with the issuance of a driver's license). In writing for the Court, Justice Antonin Scalia noted that "neither as originally enacted nor as later amended does Title VI display an intent to create a free standing private right of action to enforce [these] regulations." Scalia explicitly stated that neither section 601 nor section 602 of Title VI permits individuals to sue on the basis of disproportionate exposure to hazardous materials (*Alexander v. Sandoval*, 532 U.S. 275 [2001]).

The Supreme Court's decision has been devastating to the environmental justice movement. As one legal expert has observed,

> The effect of Justice Scalia's decision in Sandoval is very profound. For over fifty years, Civil Rights advocates challenged agency decisions using § 602 [of Title VI]. There was extensive case law, which established that an implied private right of action existed under § 602. Furthermore, Justice Scalia indicated § 1983 may not be used to enforce a violation of § 602 because § 601 does not prohibit unintentional discrimination. The only way advocates can challenge agency decisions is to prove intentional discrimination. Requiring plaintiffs to prove intentional discrimination to substantiate disparate impact claims, without alternate legal mechanisms, will sound the death knell for the environmental justice movement (Zywicki 2002).

Other activists are more sanguine. However dismal the situation may look at the federal level, state courts operating under state constitutions may be more receptive to claims for redress because of environmental inequities. In California, for example, recent changes in the law have made it permissible for individuals to sue state agencies to enforce state regulations. As one legal expert has written, such changes may mean that "California may soon emerge as the leader in the effective use of civil rights laws to address environmental inequities" (Rechtschaffen 2003).

The U.S. Congress Has Failed to Take Action on Basic Environmental Justice Legislation

The first legislation dealing with environmental justice in the U.S. Congress was introduced by Rep. John Lewis (D-GA) in 1992. That bill was referred to the House Subcommittee on Water Resources and Environment, but no action was ever taken. Lewis reintroduced the bill in 1998, and it was referred to the Subcommittee on National Parks and Public Lands. Again, no action was taken. Similar legislation on environmental justice was introduced into later sessions of Congress, with similar bills appearing, in some cases, in both the House of Representatives and Senate. In no case did a bill ever get any further than being referred to some subcommittee of the House or Senate.

After President Bill Clinton's Executive Order 12898 of 1994, the purpose of these bills was to codify into law the provisions of the executive order. While members of the government bureaucracy are *supposed* to abide by the provisions of an executive order, they are not constrained to do so in the same way they would be by a law passed by Congress. Supporters of environmental justice have, therefore, argued for more than 15 years that Congress should pass a law to strengthen the requirements of Clinton's executive order. A reading of the most recent versions of the environmental justice bills shows how similar they are to the executive order itself.

Given that national legislators have clearly decided not to act on environmental justice bills, it appears that the movement cannot hope, for the foreseeable future, for support for its cause from the U.S. House or Senate. As noted above, advocates have decided, therefore, to look to state legislatures for such actions. A number of states have moved in this direction, with laws on environmental justice now having been passed in about a dozen states. These laws vary widely from state to state, with some establishing a complete set of environmental justice provisions similar to those found in Executive Order 12898 and others dealing with only a specific aspect of environmental justice. For example, the state of Oregon adopted new environmental justice legislation in 2007, which took effect on January 1, 2008, requiring all 14 state agencies that deal with natural resources to develop procedures that will increase the participation of citizens affected by agency actions. The law also created an environmental justice task force charged with advising the governor and natural resources agencies on environmental justice issues, with defining environmental justice issues in the state, with identifying minority and low-income communities that may be affected by environmental decisions made by the agencies, with meeting with environmental justice communities, and with making recommendations to the governor regarding concerns raised by these communities.

Other state laws are more limited. Mississippi, for example, has an "anti-concentration" law that prohibits the siting of more than one hazardous facility in any one county of the state. The law contains a loophole, however, as it allows a variance from this restriction if state needs change in such a way as to make the construction of more than one facility advantageous. Alabama and Georgia have similar anti-concentration laws. An example of some of the weakest environmental justice legislation is North Carolina's law on the issuance of solid waste permits. The law requires a permitting agency to "consider" the most recent census data for an area in which a waste disposal site is to be located, but it gives no guidance as to how the agency is to use that data.

Environmental Justice Activists Have Learned to Use a Combination of Tools for Achieving Their Goals

Any organization striving to eliminate environmental inequities will necessarily have to consider legal and administrative remedies for such situations. But given the poor record of federal legal and judicial bodies in responding to environmental justice claims, activists have long realized that they also need to explore other tools for reaching their objective. These tools include the intensive involvement of community members; reliance on experts in science, technology, the law, sociology, political science, and other fields; networking among social, civil rights, and environmental justice organizations; and the use of education and political pressure on decision makers. An example that illustrates the way environmental justice activists can use these tools is the Environmental Justice for Cleaning Workers Campaign (EJCWC) of the Massachusetts Coalition for Occupational Safety and Health (MassCOSH). MassCOSH is a nonprofit organization of workers, unions, community groups, and health and safety activists to organize and advocate for safe, secure jobs and healthy communities throughout eastern and central Massachusetts.

The focus of the EJCWC is the health of more than 30,000 custodians and janitors who work in homes, commercial buildings, and a variety of public facilities throughout the commonwealth of Massachusetts. These workers wash and strip floors, clean toilets and other equipment, maintain and shine furniture, and carry out a variety of other tasks that require the use of materials that may be irritating or hazardous to one's health. According to some evidence, these materials may be implicated in high rates of asthma, bladder and prostate tumors, heart attack deaths, cancer of the lower urinary tract, chemical burns, stillbirths and preterm delivery, birth defects, and Down syndrome.

The vast majority of cleaning workers in Massachusetts (as is the case throughout the country) are people of color and/or immigrants. These workers seldom have the education or experience that allows them to understand the hazards they face or to take action to remove such hazards from their workplace.

MassCOSH has decided on a three-tier program to improve the working conditions for cleaning workers in Massachusetts. The first tier involves active participation by workers themselves to collect the information needed to document the hazards they face on the job. Tier two calls for organizing workers and supporters of EJCWC to promote safer working conditions. Tier three involves public advocacy among decision makers to replace hazardous working materials with green products that do not pose a health risk for workers. To achieve these goals, MassCOSH formed a partnership with other stakeholder organizations in Massachusetts, such as the Service Employees International Union (SEIU) Local 615 and its nonprofit subsidiary, the Voice and Future Fund; the Boston School Custodial Union; the Boston Urban Asthma Coalition (BUAC); the Boston Urban Asthma Initiative; and Urban Edge and Viet AID, two local community development corporations.

By following the above agenda, MassCOSH has achieved remarkable success in reducing the environmental hazards to which cleaning workers are exposed. Data collected from workers in 2004 have led some major public employers to change the way their facilities are cleaned. The contractor responsible for cleaning Logan International Airport, for example, has agreed to replace all hazardous cleaning materials with safer alternatives and to provide a display at the airport explaining to workers how the new materials are to be used. SEIU Local 615 has integrated health and safety materials into its training manual for shop stewards. And the Massachusetts Bay Transportation Authority (MBTA) has agreed to use green cleaning products in place of hazardous materials.

In 2005, custodians in four of Boston's public schools conducted a pilot program to study the effect on asthma rates among employees using green cleaning materials rather than traditional cleaning supplies. As a result of that study, the Boston school system announced in 2006 that it would begin to change its cleaning procedures and start purchasing only green products in the future. Also in 2005, MassCOSH responded to a request from Urban Edge to analyze and reassess its cleaning program. Urban Edge is a community development corporation that works for the development of sustainable housing in the Jamaica Plains and Roxbury neighborhoods of Boston. The two entities were able to develop a new set of cleaning guidelines that replaced materials and practices hazardous to a worker's health with green products that are safer to use. (For a copy of these guidelines, see http://urbanedge.org/green-housing.php?subcode=Green%20Cleaning.) The combined efforts of MassCOSH and its collaborators have demonstrated how proponents of environmental justice can bring about significant change, even absent the force of federal or state legislation and/or the power of judicial decisions.

The battle of the Sokaogon Mole Lake Chippewa and Forest County Potawatomi Indian tribes in Wisconsin to prevent development of a zinc and copper mine near their land provides another illustration of imaginative ways in which environmental justice activists can achieve success in their efforts. The battle goes back to the 1970s, when the Exxon Corporation purchased more than 5,000 acres of land about 100 miles north of Green Bay. Over time, ownership changed hands a number of times, with Nicolet Mineral Company being the most recent owner.

From the moment the mine was originally proposed, members of the two tribes expressed concerns about possible environmental effects on the health of their members. They feared that runoff from mining operations would contaminate the wild rice beds upon which they depend for an important part of their food supply. They also worried that wastes from the mine would pollute their water supplies and contaminate the air with particulates. For more than 25 years, however, the tribes were unsuccessful in deterring Nicolet and its predecessors from proceeding with plans for the mine—until October 2003. At that point, an extended period of bargaining between the two tribes and Northern Wisconsin Resources Group, the timber arm of Nicolet, concluded with an agreement to sell the 5,939-acre mine site, the mineral rights, and some associated timber rights to the two tribes for $16.5 million. About half of the money was to come from profits from a Potawatomi gaming casino near Milwaukee, and the rest from a loan to be obtained by the Mole Lake tribe.

When the deal was announced, representatives of the tribe also reported that they would immediately withdraw a pending application that had been filed by the Nicolet company for a permit to mine in the area. The tribes had finally resolved a very long controversy over operations they believed would threaten their health.

"Buying off" a corporation whose activities pose an environmental risk is only one more approach to the resolution of environmental justice issues. Lacking access to administrative, legislative, and judicial venues for such cases, environmental justice activists are becoming increasingly skillful and effective in developing and using the tools that remain available to them to obtain relief from the problems they face.

References

Amer, Mildred L. 2008. *Women in the United States Congress: 1917–2008*. Washington, DC: Congressional Research Service. Updated May 7.

Anderton, Douglas L., et al. 1994a. "Hazardous Waste Facilities: 'Environmental Equity' Issues in Metropolitan Areas." *Evaluation Review* 18 (2): 123–140.

Anderton, Douglas L., Andy B. Anderson, John Michael Oakes, and Michael R. Fraser. 1994b. "Environmental Equity: The Demographics of Dumping." *Demography* 31:229–248.

Anderton, Douglas L., John Michael Oakes, and Karla L. Eagan. 1996. "Methodological Issues in the Spatiotemporal Analysis of Environmental Equity." *Social Science Quarterly* 77:508–515.

Anderton, Douglas L., John Michael Oakes, and Karla L. Egan. 1997. "Environmental Equity in Superfund: Demographics of the Discovery and Prioritization of Abandoned Toxic Sites." *Evaluation Review* 21 (1): 3–26.

Austin, Regina, and Michael Schill. 1994. "Black, Brown, Red, and Poisoned." *The Humanist* 54 (July/August): 9–16.

Bailey, Conner, and Charles E. Faupel. 1992. "Environmentalism and Civil Rights in Sumter County, Alabama." In *Race and the Incidence of Environmental Hazards: A Time for Discourse*, edited by Bunyan Bryant and Paul Mohai, chap. 11. Boulder, CO: Westview Press.

Been, Vicki. 1994. "Locally Undesirable Land Uses in Minority Neighborhoods: Disproportionate Siting or Market Dynamics?" *Yale Law Review* 103(6): 1383–1422.

Binkly, Gail. 2007 (November 8). "'Economic Blackmail' Fueling Power-Plant Push? 'Big Coal' Author Says Desert Rock Project Makes No Sense in Southwest." [Online article or information; retrieved 5/24/08] Alpcentral. http://www.alpcentral.com/article.php?article_id=1948.

Boerner, Christopher, and Thomas Lambert. 1995. "Environmental Justice Can Be Achieved through Negotiated Compensation." In *Environmental Justice*, edited by J. S. Petrikin, chap. 7. San Diego: Greenhaven Press.

Bonorris, Steven. 2004. *Environmental Justice for All: A Fifty-State Survey of Legislation, Policies, and Initiatives.* San Francisco: Hastings College of the Law, University of California.

Bullard, Robert D. 1983. "Solid Waste Sites and the Black Houston Community." *Sociological Inquiry* 53 (2–3): 273–288.

Bullard, Robert D. 1992. "Environmental Blackmail in Minority Communities." In *Race and the Incidence of Environmental Hazards: A Time for Discourse,* edited by Bunyan Bryant and Paul Mohai, chap. 6. Boulder, CO: Westview Press.

Bullard, Robert D. 1994. *Worst Things First? The Debate over Risk-Based National Environmental Priorities.* Washington, DC: Resources for the Future.

Bullard, Robert D. 2000. *Dumping in Dixie: Race, Class, and Environmental Quality,* 3rd ed. Boulder, CO: Westview Press.

Bullard, Robert D., Paul Mohai, Robin Saha, and Beverly Wright. 2007. *Toxic Wastes and Race at Twenty: A Report Prepared for the United Church of Christ Justice & Witness Ministries.* Cleveland, OH: Justice and Witness Ministries, United Church of Christ.

Buttel, Frederick R. 1987. "New Directions in Environmental Sociology." *Annual Review of Sociology* 13:465–488.

Davidson, Pamela, and Douglas L. Anderton. 2000. "Demographics of Dumping II: A National Environmental Equity Survey and the Distribution of Hazardous Materials Handlers." *Demography* 37 (4): 461–466.

Davis, M. E. 1977. "Occupational Hazards and Black Workers." *Urban Health* 6 (August): 16–18.

"Environmental Justice Executive Order [New Jersey]." 2004. [Online article or information; retrieved 5/24/08.] http://www.state.nj.us/dep/ej/ejeo.pdf.

Featherstone, Liza. 2005 (August 1). "E-Raced: EPA Says Race, Income Shouldn't Be Environmental Justice Factors." Grist. [Online article or information; retrieved 6/19/08.] http://www.grist.org/news/maindish/2005/08/01/featherstone-ej/.

Gelobter, Michel. 1992. "Toward a Model of 'Environmental Discrimination.'" In *Race and the Incidence of Environmental Hazards: A Time for Discourse,* edited by Bunyan Bryant and Paul Mohai, chap. 5. Boulder, CO: Westview Press.

Gerrard, Michael B. 2003. "EPA Dismissal of Civil Rights Complaints." *New York Law Journal,* November 28, 3+.

Gianessi, Leonard P., Henry M. Peskin, and Edward Wolff. 1979. "The Distributional Effects of Uniform Air Pollution Policy in the United States." *Journal of Economics* 93 (May): 281–301.

Gogal, Daniel. 2008. Personal communication, April 29.

Goldman, Benjamin A. 1993. *Not Just Prosperity: Achieving Sustainability with Environmental Justice.* Washington, DC: National Wildlife Federation.

Hamilton, James T. 1993. "Politics and Social Cost: Estimating the Impact of Collective Action on Hazardous Waste Facilities." *RAND Journal of Economics* 24 (Spring): 101–125.

Infoplease. 2008 (June 19). "The Wage Gap, by Gender and Race." [Online article or information; retrieved 6/19/08.] http://www.infoplease .com/ipa/A0882775.html.

Johnson, F. Reed. 1980. "Income Distributional Effects of Air Pollution Abatement: A General Equilibrium Approach." *Atlantic Economic Journal* 8 (4): 10–21.

Jones, Robert Emmet, and Shirley A. Rainey. 2006. "Examining Linkages between Race, Environmental Concern, Health, and Justice in a Highly Polluted Community of Color." *Journal of Black Studies* 36 (4): 473–496.

Kazis, Richard, and Richard L. Grossman. 1982. *Fear at Work: Job Blackmail, Labor, and the Environment.* New York: Pilgrim Press.

Lavelle, Marianne, and Marcia Coyle. 1993. "Unequal Protection: The Racial Divide in Environmental Law." *National Law Journal* 21 (September): S1–S12.

Louisiana Advisory Committee to the U.S. Commission on Civil Rights. 1993. *The Battle for Environmental Justice in Louisiana . . . Government, Industry, and the People.* Kansas City, MO: U.S. Commission on Civil Rights, Central Regional Office.

Lowe, G. D., and T. K. Pinhey. 1982. "Rural-Urban Differences in Support for Environmental Protection." *Rural Sociology* 47 (1): 114–128.

Mohai, Paul. 2003. "Dispelling Old Myths: African American Concern for the Environment." *Environment Magazine* 45 (June): 10–26.

Mohai, Paul, and Bunyan Bryant. 1992. "Environmental Racism: Reviewing the Evidence." In *Race and the Incidence of Environmental*

Hazards: A Time for Discourse, edited by Bunyan Bryant and Paul Mohai, chap. 13. Boulder, CO: Westview Press.

National Academy of Public Administration. 2002. *Models for Change: Efforts by Four States to Address Environmental Justice.* Washington, DC: National Academy of Public Administration.

New York State Department of Environmental Conservation. 2003. "Environmental Justice and DEC Permits. Summary of Commissioner Policy CP-29, Environmental Justice and Permitting." [Online article or information; retrieved 5/24/08.] http://www.dec.ny.gov/public/36929.html. Latest Date Revised: March 19, 2003.

Office of the Inspector General. Environmental Protection Agency. 2004. *EPA Needs to Consistently Implement the Intent of the Executive Order on Environmental Justice.* Washington, DC: Environmental Protection Agency.

Office of the Inspector General. Environmental Protection Agency. 2006. *EPA Needs to Conduct Environmental Justice Reviews of Its Programs, Policies, and Activities.* Washington, DC: Environmental Protection Agency.

Pace, David. 2005 (December 13). "AP: More Blacks Live with Pollution." [Online article or information; retrieved 5/24/08.] http://hosted.ap.org/specials/interactives/archive/pollution/part1.html.

Peter, Ellen M. 2001. "Implementing Environmental Justice: The New Agenda for California State Agencies." *Golden Gate University Law Review* 31 (Spring): 529–591.

Pierce, Margo. 2007 (April 8). "Equal Environmental Rights: Ending the Link between Poverty, Race and Pollution." *City Beat.* [Online article or information; retrieved 5/24/08.] http://citybeat.com/2007-04-18/cover.shtml.

Probst, Kate. 1990. *Hazardous Waste Sites and the Rural Poor: A Preliminary Assessment.* Arlington, VA: Clean Sites, Inc.

Rechtschaffen, Clifford. 2003. "Using California's Anti-Discrimination Law to Remedy Environmental Injustice." Environmental Justice Committee Newsletter, ABA Section of Individual Rights and Responsibilities. http://www.abanet.org/irr/committees/environmental/newsletter/dec03/Civilrights.html.

Sayen, Jamie. 1995. "Obligatory Skepticism: The Environmental Movement's Internal Enemy." *Earth Island Journal* 10 (Spring). [Online article or information; retrieved May 28, 2008.] http://www.earthisland.org/eijournal/new_articles.cfm?articleID=830&journalID=75.

Taylor, Dorceta E. 1989. "Blacks and the Environment: Toward an Explanation of the Concern and Action Gap between Blacks and Whites." *Environment and Behavior* 21 (March): 175–205.

Toffolon-Weiss, Melissa, and J. Timmons Roberts. 2004. "Toxic Torts, Public Interest Law, and Environmental Justice: Evidence from Louisiana." *Law & Policy* 26 (April): 259–287.

U.S. Commission on Civil Rights. 2003. *Not in My Backyard: Executive Order 12898 and Title VI as Tools for Achieving Environmental Justice.* Washington, DC: U.S. Commission on Civil Rights.

U.S. Congress. 1993. House. Committee on the Judiciary. *Environmental Justice. Environmental Justice Hearings before the Subcommittee on Civil and Constitutional Rights,* 103rd Congress, 1st sess., March 3 and 4. Washington, DC: Government Printing Office.

U.S. Environmental Protection Agency (EPA). 1992. *Women, Minorities and People with Disabilities.* Washington, DC: Environmental Protection Agency.

U.S. Environmental Protection Agency (EPA). "Environmental Justice." [Online article or information; retrieved 5/24/08.] http://www.epa.gov/compliance/environmentaljustice/. Updated February 5, 2008.

U.S. Environmental Protection Agency (EPA). "Environmental Justice Bibliography Database." [Online article or information; retrieved 5/24/08.] http://cfpub.epa.gov/ejbib/. Updated June 18, 2006.

U.S. Government Accountability Office (GAO). 2005. *Environmental Justice: EPA Should Devote More Attention to Environmental Justice When Developing Clean Air Rules.* Washington, DC: U.S. Government Accountability Office.

Whittaker, Matthew, Gary Segura, and Shaun Bowler. 2005. "Racial/Ethnic Group Attitudes toward Environmental Protection in California: Is 'Environmentalism' Still a White Phenomenon?" *Political Research Quarterly* 58 (3): 435–447.

Zywicki, Michael. 2002. "Scalia Scalds Environmental Justice Movement." *Vermont Journal of Environmental Law* 4 (October 10). [Editorial.] [Online article or information; retrieved 5/24/08.] http://www.vjel.org/editorials/ED10033.html#ref20.

3

Worldwide Perspective

The environmental justice movement in the United States is premised on the belief that low-income and minority communities are exposed to a disproportionate share of the hazardous waste products (in one form or another) produced by the nation's industries. Study after study conducted over the past three decades has demonstrated the validity of that premise. In recent years, a similar pattern on an international scale has become apparent. At least partly because of the globalization of business, raw materials and waste products now flow more freely than ever within and among countries and throughout the world. As a result of that process, low-income communities in developing countries around the world are increasingly exposed to the hazards involved in the production, consumption, and disposal of a seemingly limitless range of products, from pesticides and drugs to nuclear power plants and electricity generating facilities. This chapter considers some examples of environmental racism throughout the world, the special problems posed by global climate change and the need for sustainable communities, and the progress being made by environmental justice groups worldwide.

Environmental Inequities Worldwide

As the U.S. example shows, environmental inequities can occur in any country, whether highly developed or still in its earliest stages of development. On an international scale, however, one overall pattern tends to predominate: Developed nations such as the United States and members of the European Union tend to treat less-developed nations in much the same way that industries in the United States treat low-income and minority communities. That is, they place resource extraction facilities, garbage dumps, industrial plants, and other polluting operations in areas where residents are generally less able to object to the environmental insults associated with these operations. The following cases illustrate only a few of the ways in which this generally north-to-south environmental injustice has occurred in the past.

Mineral Extraction in Papua New Guinea

One might easily think of the island of New Guinea as a somewhat backward region of spectacularly beautiful scenery having but little contact with the modern world. That picture is only partially accurate. The scenery is spectacular, containing a number of the rarest animal and plant species in the world, but modern industrialism has become an important part of Papuan culture.

Starting in the 1950s, the Australian government (which then governed Papua) sold mining rights on the island to multinational corporations with no input from local citizens or their leaders and essentially no concerns about possible deleterious impacts on the local environment or residents' health and welfare. When Papua New Guinea (PNG) achieved its independence in 1975, the new government continued that practice. As a desperately poor nation, PNG saw resource extraction as a critical way of attracting foreign investment in the young country.

As is sometimes the case, mining companies paid relatively little attention to laws and regulations relating to environmental protections. Indeed, the PNG government even exempted some corporations from its own environmental laws, accepting environmental impact statements written by the companies as being more "suitable" than those of its own government agencies.

One of the most notorious cases of environmental injustice in PNG involved a mine at Ok Tedi, near the mouth of the Fly River, that belonged to the Australian corporation BHP-Billiton. For the two decades during which it operated, the mine produced an average each day of about 80,000 tons of tailings containing copper, zinc, cadmium, and lead, all toxic to plant and animal life (Marshall 2002). Before long, the mine's toxic wastes had decimated plant and animal life along the Fly River. For residents of the region, the most serious problem was the loss of the abundant fish population, which formed a major part of their diet. But one study showed that, beyond the loss of fish, "[a] decline in the turtle population has also been seen as the sandy banks where eggs were laid have been replaced by particulate copper and mud. Prawns, lobsters, and bivalves have also declined, and birds that depend on riverine life (egrets, kingfishers, Brahmin kites) have also left. The water is no longer safe for bathing, washing clothes, swimming, or drinking" (Johnston 1994, 92). Over time, the potential liabilities for BHP continued to grow until, in January 2002, it negotiated a deal with PNG to sell the mine to the government. In return, the government was to indemnify the company against all future claims.

As a result of that agreement, BHP was able to exit PNG facing no liabilities for its actions in the country. The government took over ownership of the mine, which it decided to keep operating for at least a decade because of its enormous potential for income. And the people of PNG, who had already suffered severely from environmental degradation, faced another 10 years of mine operation with its consequent destruction of their lands.

Throughout this sordid affair, evidence accumulated revealing PNG governmental and BHP corporate dismissive attitudes toward indigenous peoples of the nation. They were apparently heard to comment that those people were not intelligent enough to understand what was going on in their land or to make decisions about how the land was to be used. As the sale of the mine was concluded, one representative of the PNG Environmental Law Centre said: "Based on my conversations with the local people, it is quite clear that in my opinion the people who are signing this agreement do not know what they are signing" (Marshall 2002). Perhaps not, but they had another decade to live with its consequences.

Environmental Discrimination against the Roma in Central Europe

Discussions of environmental inequities around the world often focus on obvious candidate countries in Africa, South America, and parts of Asia. But environmental discrimination also occurs in regions more commonly thought of as "developed" or "advanced." For example, many countries in Central and Eastern

Europe, some of which were part of the former Soviet bloc, have made impressive economic progress since the collapse of the Soviet Union in 1991. Still, as in the United States, it is not difficult to find examples of environmental racism in the most highly developed nations.

One recurring example involves the Roma (or Romany), also known as gypsies, who occupy most of Central Europe. Historically, the Roma have been subjected to widespread and aggressive discrimination in almost every nation they inhabit. In recent decades, that discrimination has shown up over and over again in the form of environmental racism.

A common theme appears in most of these cases: A company or local government seeks a site for a new factory, municipal waste dump, or other potentially hazardous entity. Because land on which the Roma live is often the least expensive and least desirable for residential purposes, it is often selected for the new facility. Roma are moved out of the area or, more commonly, allowed to remain on land immediately adjacent to the facility. Over succeeding years, a variety of health problems develop among the Roma who are exposed to the hazardous facility. These problems are exacerbated because often the only jobs available to Roma workers are at the same hazardous facility.

For example, scientists have studied in some detail a case in Heves, Hungary, where a primary source of income for Roma men was the recycling of lead taken from batteries in an illegal dump near their homes. Eventually, more than 1,500 individuals were exposed to high levels of lead, one 15-month-old girl died, and 65 children were hospitalized (Steger 2007, 21).

The situation for Roma in Romania is no better. A recent report by the Central European University Center for Environmental Policy and Law, the Health and Environment Alliance, and the Coalition for Environmental Justice listed a number of instances in which Roma were relocated by local, regional, and federal governments into areas where they would be exposed to disproportionately high levels of toxic emissions. The town of Episcopia Bihor, for example, relocated its Roma community to a new settlement that was built on top of the town's municipal rubbish dump. Under such circumstances, an epidemic of health problems is almost guaranteed. In some parts of Central Europe—Hungary, for example—Roma have a life expectancy 10 to 15 years less than comparable non-Roma populations (Steger 2007, 25–26).

Ecotourism and the Maasai

Instances of environmental injustice sometimes arise in the most unlikely of situations. Consider the Maasai (or Masai) tribe that once populated wide stretches of East Africa. The Maasai are *transhumance pastoralists*, a term referring to a population that moves about the countryside annually with its animals, constantly looking for new grazing grounds and permitting overgrazed areas to recover. With the arrival of British colonists in East Africa in the early 20th century, the Maasai were restricted to ever-smaller areas on which to live. The British saw the Maasai as a "nuisance" and negotiated more and more restrictive treaties limiting their movement to smaller and smaller regions (Narimatsu 2008).

This pattern continued throughout the 20th century, even after the replacement of British rulers by native governments beginning in the 1960s. Most young African governments realized early on the potential value of safari travel, ecotourism, and other travel experiences that emphasized the enormous natural wonders of their nations. The government of Kenya, for example, established a goal of hosting 1 million tourists by the year 1990.

One element in such programs was the creation of new national parks, a goal that, in turn, required governments to even further restrict the Maasai. Ironically, then, the effort of African governments to save endangered species such as the tiger, lion, rhinoceros, and hippopotamus resulted in severely changing the lifestyle of one of the continent's indigenous peoples. In their view, "The Maasai people have been systematically driven off our traditional lands in the wake of economic development, irresponsible tourism, large-scale farming, and other activities that destroy wildlife habitat" (Maasai Environmental Resource Coalition 2008).

For almost a century, the Maasai were essentially pawns in the economic and environmental policies of, first, the British colonists and, later, native African governments in Kenya, Tanganyika, and other East African countries. They had little experience and few tools for dealing with the more sophisticated European and native governments. Toward the end of the 20th century, however, the Maasai began to better understand the options available to them in protecting their traditional culture and began organizing to make their needs and wishes known. Maasai activists have now organized into an umbrella organization known as the Maasai Environmental Resource Coalition, consisting of about a dozen local groups. These groups include the Aramat Association, Arusha Women's Group, and Ilkisongo Maasai Cultural Organization of Tanzania; the Enkong'u Narok Cultural Center, Kajiado Integral Rural Association for Development, Maasai Mara Women's Group, and Olchorro-Oirowua Wildlife Conservation Association of Kenya; and the Maasai Cultural, Wildlife and Ethical Tourism Society of Kenya and Tanzania. The mission of the organization is to protect traditional land rights of the Maasai people and to help conserve, manage, and develop sustainable use systems for the ecosystems of East Africa.

The Bakun Hydroelectric Project in Sarawak

In the most egregious examples of environmental racism, human populations are sometimes treated as disposable commodities to be relocated and moved around without regard to their own culture, traditions, lifestyles, and best interests. A striking example of that policy can be found in plans for building the Bakun hydroelectric facility in the Malaysian state of Sarawak on the island of Borneo. The earliest studies for such a facility began in the 1960s and were approved by the Malaysian government in 1986. Four years later, however, the government changed its mind and put construction plans on hold.

In 1993, plans for the dam were revived, and a no-bid contract was signed with Ekran Berhad, a close friend of then–prime minister Mahathir Mohammad. Those plans called for the flooding of an area of about 270 square miles (about the size of the island of Singapore) at a cost of about $6 billion by the year 2003. Construction would require the removal of about 9,000 indigenous people, members of the Iban, Dayak, and Bidayuh tribes. These peoples have traditionally practiced subsistence agriculture, depending on the forest for most of their food, housing, clothing, and other needs. The area to be flooded was also home to one of the largest collections of threatened and endangered species of plants and animals in the world (Friends of the Earth 1996; Malaysian Timber Council 2002).

The dam project was halted a second time in 1997, this time because of the widespread Asian financial crisis and reduced estimates of the amount of electricity needed in Malaysia. Studies showed that energy consumption in the whole state of Sarawak was about 500 megawatts and that generating facilities already in operation were producing 700 megawatts. In addition, experts predicted that existing facilities could be expanded to produce about 1,000 megawatts. Other studies showed that energy use in Sarawak was expected to decline from 9.8 percent in 2000 to 5.6 percent in 2010. Under those circumstances, it was no longer clear that the Bakun dam was needed.

Nonetheless, plans for the dam were revived yet again in May 2000 with a new completion date of 2010. Statistics associated with the project are impressive. The dam will be the tallest concrete-faced rockfill dam in the world, capturing the largest lake in Malaysia by storage volume and producing the largest amount of electric power of any station in the country. In addition, the submarine cable that carries power from Sarawak to the Malaysian peninsula will be the longest in the world.

The required displacement of indigenous peoples has now been completed. Most of them have been relocated to a longhouse settlement in Bakun, a dramatically different lifestyle than the one with which they are familiar. Jobs at the dam that were promised them have largely not materialized, and unemployment rates are very high among the tribespeople. The deportees are expected to pay up to $15,000 for the new housing they are being provided, although such sums are essentially unimaginable to people who had previously had virtually no experience with money. Unlike some environmental justice stories, the Bakun dam saga appears to have no happy ending associated with it. The Iban, Dayak, and Bidayuh have lost their traditional homes to a giant modern industrial project.

Thor Chemicals and Mercury "Recycling"

Historians of the international environmental justice movement sometimes point to the Thor case as one of the first events to attract worldwide attention. It began in 1988, when the British company Thor Chemicals moved its mercury reprocessing operations from Margate, in Kent, England, to the small industrial village of Cato-Ridge in the South African province of Kwazulu-Natal. Thor's decision to move from England to South Africa was occasioned by concerns raised by the British government's Health and Safety Executive about excessive levels of mercury in the air around the Margate facility and in the urine of workers employed at the plant. Thor was aware that environmental protections in South Africa were less stringent than in the United Kingdom and that continued operation of the plant was possible only if the move was made.

The new Thor plant in Cato-Ridge soon drew the attention of manufacturers around the world. It was one of the few facilities worldwide willing and able to accept wastes containing mercury for reprocessing. Two of the largest contributors to the Thor recycling program were the American companies Borden Chemicals and Plastics and American Cyanamid. In a 1999 report, the U.S. Environmental Protection Agency (EPA) said that the two companies were shipping about 190 tons of mercury-containing wastes annually to the Cato-Ridge facility (Fondaw 2008).

Evidence of the harmful effects of the mercury reprocessing facility began to appear shortly after the plant began operations. Water samples taken from the Mngeweni River, into which Thor discharged its wastes, contained 1.5 million parts per billion (ppb) of mercury, 1,500 times greater than the current EPA recommended limit for mercury in water. The river is the main source of drinking water for the city of Durban (current population: 3.5 million) and the primary source of water for the region's agricultural and dairy industries. Even 40 miles downstream from the plant mercury levels reached 20 times the EPA recommended limits.

The South African government did not appear to be unduly concerned about the potential health hazards posed by the Thor plant. It ordered a four-week closure to permit the plant to clean up its operations, then issued a new license to the facility. In 1994, the Department of Environmental Affairs sent a secret memo to the South African cabinet praising Thor's "sensible operations" and commending its "sound work" (Cook and Koch 1992, 9).

The situation at Cato-Ridge first drew international attention in 1990, when the *St. Louis Post-Dispatch* conducted a study of hazardous waste disposal policies and practices in the United States. The paper traced the link from Borden Chemicals and American Cyanamid to Thor and followed up with an extensive report on conditions in and around the Thor plant. Its story soon attracted the attention of environmental groups, particularly Earthlife Africa and Greenpeace International, which began their own investigations of Thor and of the South African government's "cozy" relationship with the company. Almost simultaneously, reports of the first deaths of workers at the Thor plant began to surface. According to one report, workers at the Thor plant had mercury levels as high as 600 to 1,000 ppb in their urine, compared to generally accepted limits of 50 ppb.

As more and more information about Thor's operations became available, a number of individuals began filing suits for damage as a result of their employment at the Cato-Ridge facility. The company settled out of court for a total of $2.1 million for 20 such claims in 1997 and then again in 2000 for a total of $353,000 for an additional 20 employees. In 1992, the company announced that the plant would cease operations as of 1996. In 2000, the South African government invited experts from the EPA to visit the Thor site and recommend procedures for cleaning it up (Knight 2001).

The Voyage of the *Khian Sea*

Arguably, the most famous example of the use of developing nations for the disposal of hazardous wastes involves the voyage of the freighter *Khian Sea*, which was attempting to dispose of about 14,000 tons of ash from a municipal incinerator in the city of Philadelphia's Roxborough neighborhood. The incinerator, like most municipal incinerators, burned household wastes consisting of everything from newspapers and garbage to cigarettes and dead batteries. Lacking space within the city limits to dispose of incinerator ash, Philadelphia contracted with outlying areas (primarily in New Jersey) to accept its wastes. Although the materials burned in the Roxborough incinerator were relatively benign products of home and municipal use, the ashes that resulted were not. They contained by-products such as dioxins, furans, cadmium, and lead, considered to be toxic, teratogenic, carcinogenic, or otherwise hazardous to human health.

In 1984, the state of New Jersey informed authorities in Philadelphia that it would no longer accept the city's incinerator ash for disposal. Instead, the city contracted with the firm Paolino and Sons, Amalgamated Shipping and Coastal Carrier, to dispose of the ash. Paolino's original intent was to dispose of the wastes on a manmade island in the Bahamas—except that the Bahamian government denied the company a permit to unload its cargo on the island. Over the next year, the *Khian Sea* roamed the Caribbean looking for a place to dispose of the ash. Between September 1986 and August 1987, it was turned away from Bermuda, the Dominican Republic, Honduras, Guinea-Bissau, and the Netherlands Antilles. Efforts were then made to convince the Panamanian government to accept the ash for use in road-building in Panama, an offer that was refused (Schwartz 2000).

On December 31, 1987, the *Khian Sea* arrived in Haiti and obtained a permit from the government to unload a portion of its cargo, described to officials as "fertilizer." Between January 20 and January 31, the ship offloaded 4,000 tons of the ash, only to be interrupted by an order from the Haitian government to reload its cargo and return to sea. The *Khian Sea* did return to sea, but without reloading the ash it had already left on Gonaives beach.

After a brief return to Philadelphia in February 1988, the *Khian Sea* set off once more to find a dumping ground for the remaining ash in its hold. Over the next six months, it was refused permission to unload its cargo in Morocco, Senegal, Sri Lanka, and Yugoslavia, where it put in for repairs in August 1988. When it was escorted from the country by the Yugoslav navy, it had a new name—the *Felicia*—and a new country of registry—Honduras. Before long, its name was changed once more, this time to the *Pelecano*. On September 28, 1988, it passed through the Suez Canal, listing Singapore as its destination.

By the time the *Khian Sea/Felicia/Pelecano* reached Singapore in November 1988, the troublesome ash cargo had disappeared. The captain later admitted that the ash had been dumped into the Indian and Atlantic oceans on the way from North America to the Pacific. In 1993, William P. Reilly and John Patrick Dowd, officers of Coastal Carriers Inc., which operated the *Khian Sea*, were convicted of perjury in the United States and sentenced to jail. No convictions were ever obtained for the dumping of ash in Haiti or on the open seas.

In 1997, a coalition of the Greenpeace environmental organization and Haitian environmental groups joined to form the Return to Sender project, aimed at having Philadelphia's hazardous ash removed from Gonaives beach. The project was eventually able to obtain funds from the city of Philadelphia to hire Waste Management, Inc. (WMI), to complete that task. In April 2000, WMI began the work and eventually removed the remaining 2,500 tons of waste to a docking point in Florida's St. Lucie Canal. After resting at that location for two years, the ash was returned to its final burying ground in Franklin County, Pennsylvania, by WMI in June 2002 (Greenpeace 2008).

A Background for the Transboundary Dumping of Hazardous Wastes

The case studies presented here may strike an observer as examples of horrible disregard for the health and lives of people in less-developed countries (LDCs) by uncaring corporations and upscale consumers in more-developed countries (MDCs). Although that assessment may be correct in some ways, it ignores developments in MDCs such as the United States that made these environmental injustice examples not only possible but perhaps inevitable.

One of the important but unexpected consequences of the American environmental movement of the 1960s and 1970s was that the United States began to adopt laws and regulations far more severe than those of most other nations of the world, certainly more restrictive than those of less-developed nations. Pesticides that could no longer be used in the United States because of their toxicity were, in many cases, still legal in other nations. Companies that could no longer sell their products to American farmers because of their toxicity might well be able to sell them to farmers in Indonesia, Ghana, Somalia, and other struggling nations (Norris 1982).

Such practices could be thought of as just another form of environmental inequity, and perhaps environmental racism, but on a much grander scale. During the late 1980s and early 1990s, a number of waste disposal problems made this analogy even more obvious. In 1988, for example, the city of Philadelphia hired a Norwegian company, Bulkhandling, Inc., to transport 15,000 tons of toxic incinerator ash to Kassa Island, part of the African nation of Guinea. Only when plant and animal life began to die off on the island did the government discover what had happened and ordered Bulkhandling to remove and return the ash to Philadelphia (Mpanya 1992).

As permits and land for hazardous waste landfills on the continental United States became more and more difficult to obtain, dumping in African nations became more and more attractive. The number of such cases increased by a factor of 10 in the three-year period between 1987 and 1989, rising from 3 cases in 3 countries in the first of these years to 40 cases in 25 nations in 1989 (Mpanya 1992).

As with possible cases of environmental racism in the United States, such practices often make good economic sense: They provide an answer to a difficult corporate problem (waste disposal) while bringing significant economic benefit to people who need it (poor African nations). In fact, this argument in a slightly different context was made by Lawrence Summers, chief economist of the World Bank, in a memo of December 12, 1991. The memo was obtained by and then reprinted in the British journal *The Economist* in its February 8, 1992, issue. "Just between you and me," Summers wrote, "shouldn't the World Bank be encouraging more migration of the dirty industries to the LDCs?" (The real author of the now infamous "Summers memo" was apparently one of his aides, Lant Pritchett, a longtime employee of the World Bank and, as of 2008, professor of economics at Harvard University [Wade 1997, 712].)

The Summers memo presented three reasons for this recommendation. First, it argued that the cost of illness and death resulting from pollution is essentially a measure of lost earnings from these conditions. Since wages are lowest in LDCs, the economic cost of exposure to pollution is lowest in those countries. Second, the memo suggested that the cost of pollution is nonlinear. That is, as the amount of pollution increases, the economic costs it produces grow geometrically. The countries best able to absorb the *initial* costs of pollution problems, then, are those that already have very little pollution, the LDCs. The writer of the memo said specifically, "I've always thought that under-populated countries in Africa are vastly UNDER-polluted, their air quality is probably vastly inefficiently low compared to Los Angeles or Mexico City" (Whirled Bank Group 2008).

Finally, the memo writer said that people living in LDCs probably do not view exposure to pollution in the same way as do people in MDCs. In the latter, people live relatively long lives, during the latter parts of which they develop a number of diseases that eventually cause their deaths, such as prostate cancer. But people who live in LDCs tend to have much shorter life spans. They are, therefore, less likely to worry about late-life diseases and more likely to appreciate the benefits of an enhanced income early in their lives, even if it means exposure to potentially toxic or hazardous materials. As an adjunct to this point, the memo writer pointed out that complaints about pollution often refer to "visibility impairing particulates" that may be aesthetically displeasing but that "may have very little direct health impact" (Whirled Bank Group 2008).

The Summers memo is hardly the only expression of support for the shipment of hazardous wastes from MDCs to LDCs. For example, in a 1995 book for young adults presenting a variety of viewpoints on environmental justice, Harvey Alter, manager of the Resources Policy Department of the U.S. Chamber of Commerce, argued for a more liberal view of such trade. Humanitarian instincts, he said, prompt us to want to cut back on our trade in hazardous wastes, but the long-term effect of bans on such trade would be disastrous for underdeveloped countries because it brings very poor nations a dependable and significant source of income, thereby improving the lives of their citizens. The net effect of efforts to ban trade in hazardous wastes, he said, would be to make poverty in developing nations an even more serious problem and, at the same time, to increase the severity of the environmental problems nations are trying to solve (Alter 1995).

Alter and the Chamber of Commerce have also promoted this view of trade on the international stage. During early discussions over the Basel Convention on the Control of Transboundary Movements of Hazardous Wastes and Their Disposal (the so-called "Basel Convention"; see below for more detail), Alter and the chamber argued vigorously against a proposed ban on all shipment of waste materials between Organisation for Economic Development and Co-operation (OECD) states and non-OECD nations. In fact, some observers have credited Alter and the chamber with effectively scuttling U.S. efforts to ratify the treaty (which it still has not done as of late 2008), thus preventing implementation of this aspect of the convention (Clapp 2001, 75–76).

The Basel Convention

In spite of the views of Larry Summers (and/or his aides), Harvey Alter and the U.S. Chamber of Commerce, and like-minded individuals and groups, concerns about the deleterious effects of international trade in hazardous wastes grew throughout the 1980s. That concern culminated in the adoption in 1989 of the Basel Convention on the Control of Transboundary Movements of Hazardous Wastes and Their Disposal, also known simply as the Basel Convention. The Basel Convention was signed in Basel,

Switzerland, on March 22, 1989. It was an agreement among 103 nations establishing certain requirements for the shipment of hazardous wastes between nations. In particular, it required that a nation desiring to ship hazardous wastes to a second nation notify that nation of its intent and that the receiving nation acknowledge its willingness and consent to receive the shipment.

The Basel treaty arose, in a general sense, as the result of two major changes in the disposal of hazardous wastes. The first change was the accumulation of new environmental legislation in many developed countries limiting or restricting the production, use, and disposal of toxic and hazardous materials. Industries in the United States, European Union, Canada, Australia, and other developed countries that had previously discarded their hazardous wastes in the ground, in lakes and rivers, in the oceans, or into the air could no longer use those methods of disposal. They found it necessary to ship their wastes to other parts of the world, to nations that had not yet developed stringent environmental legislation. In addition, the technical means for making such transfers became much more sophisticated, so that a U.S. industrial facility could rather easily package its wastes for shipment to almost any part of the world.

Within this general milieu, however, was a series of specific incidents involving the transfer of hazardous wastes from MDCs to LDCs that provided the impetus for the meetings that led to the Basel Convention. Perhaps the best known of these incidents was the long voyage of the freighter *Khian Sea*, described previously. A number of other similar events, however, contributed to a growing international concern about the problem of hazardous waste transfers across national boundaries.

As with many international agreements, the Basel treaty had a long gestation period. The first formal step in the development of an international agreement about trade in hazardous wastes occurred in May 1981, when the governing council of the UN Environmental Programme (UNEP), meeting in Montevideo, Uruguay, appointed the Ad Hoc Committee of Legal Experts to review the problem of international trade in hazardous wastes. The committee recommended the appointment of a working group to develop policies and plans for the handling, transport, and disposal of hazardous wastes. More than six years later, the governing council of UNEP accepted a report of the working group (known as the Cairo Guidelines) and authorized the appointment of a new working group to begin planning for an international convention on the international movement of hazardous wastes. The first steps in planning for that convention were a series of organizational meetings involving experts from 96 nations and 50 organizations held in Budapest, Hungary; Geneva (twice); Caracas, Venezuela; Luxembourg; and Basel. These organizational meetings, in turn, led finally to the Basel Convention itself, held in the Swiss city of that name in March 1989.

The Basel Convention met with mixed success. On the one hand, 35 individual nations and the European Community did sign the convention. But a number of important states, most prominently, the United States, refused to sign. The U.S. government strongly opposed any ban on the shipment of hazardous waste to developing nations even though those nations themselves strongly supported such a ban. In the end, no African government signed the convention either. The most prominent nongovernmental environmental group present at the meeting, Greenpeace International, walked out of the final meetings, claiming that the convention had become an "instrument of legalization of waste trade [because it failed] to ban even the most immoral and environmentally damaging waste trade" (Basel Action Network 2008).

As with all international treaties, signings and ratifications continued over many years. In May 1992, the convention entered into force when Australia became the 20th nation to ratify the treaty. As of late 2008, 170 nations had signed the Basel Convention, and all but a handful had also ratified the treaty. The current holdouts are Afghanistan, Haiti, and the United States. Authorities have offered a number of reasons for the unwillingness of the United States to sign the amended Basel treaty, but some believe that companies interested in continuing the trade in hazardous wastes have had an important influence on the government's decision. In the view of Jim Pluckett, coordinator of the Basel Action Network, for example, "the United States position on this important moral and environmental question has been dictated by a very small sector of the business community. This special interest lobby wishes to maintain a free trade in toxic wastes so that they can export their pollution problems to dubious, dangerous and dirty recycling operations in developing countries" (Center for International Environmental Law 2001). Reasons for the failure of Afghanistan and Haiti to sign the Basel treaty are not clear, although the disorganized state of both national governments over the past decade or more may well be a factor.

Other Treaties and Bans

Perceived deficiencies in the final Basel treaty of 1989 were serious enough that a handful of the 116 nations attending the convention refused to sign the final agreement and walked out of the meetings before they were adjourned. Feelings ran highest among African nations, which were angry that actual bans on the trade of hazardous wastes were not instituted. These nations decided to conduct a series of their own meetings on the trade in hazardous wastes. Those meetings eventually led to a number of other agreements. The first such agreement was negotiated in 1989 as part of a series of treaties between the European Union (EU) and 71 African, Caribbean, and Pacific (ACP) nations. These treaties are known as the Lomé Conventions, after the capital city of Togo, where most of the meetings were conducted. Lomé IV, concluded in December 1989, contained provisions severely limiting the shipment of hazardous wastes from the EU to any ACP country without that country's express permission.

Three other international agreements were signed in the next two years. The first was concluded in October 1990 at a meeting in Trinidad and Tobago of environment ministers from Latin America and the Caribbean. A second agreement was signed two months later by presidents of Central American countries. And the third treaty—the Bamako Convention—was adopted in January 1991, when 12 African nations meeting under the auspices of the Organization of African Unity (OAU) agreed to prohibit imports of hazardous wastes to the African continent.

Meanwhile, individual nations and regional groups had begun to pass legislation and adopt regulations related to the transboundary shipment of hazardous wastes. On February 1, 1993, for example, the EU adopted Shipping Regulation 259/93, which banned the shipment of hazardous wastes from EU member states to any country other than those that are members of the European Free Trade Association (EFTA). That ban became effective on May 1, 1994 (European Economic Community 2008). Similar bans on the shipment of hazardous wastes were recommended and/or adopted later in the same year by other regional groups: the Association of Southeast Asian Nations (ASEAN) in September, members of the Convention for the Protection of the Mediterranean Sea against Pollution (the so-called Barcelona Convention) in October, and attendees at a meeting of the Economic Commission for Latin America and the Caribbean

(ECLAC) in November. By the end of 1993, more than 100 nations had also adopted bans on the exportation and/or importation of hazardous wastes into and/or out of their own countries under at least some circumstances (Basel Action Network 2008).

Adoption of the Basel Convention has by no means solved the problem of the transboundary shipment of hazardous wastes. The original treaty was somewhat vague in many respects, and a number of meetings were held to clarify the meanings of terms and to develop clear and strong bans against the shipment of hazardous wastes to developing countries. In September 1995, the third conference of the Parties to the Basel Convention meeting in Geneva adopted an amendment to the treaty intent on clarifying and strengthening the original document's stand on hazardous wastes. A group of developed countries, including Australia, Canada, Japan, New Zealand, South Korea, and the United States, opposed the amendment, so other members of the convention decided to go ahead without the support of these nations. The conference adopted the amendment with specific reference to the group of nations most strongly in agreement with its intent: the EU, the OECD, and the principality of Liechtenstein (Basel Convention 2008). As of late 2008, 63 nations have ratified the ban amendment. As noted above, the United States has still not ratified either the original Basel Convention or the ban amendment. Although many government officials, politicians, experts, and ordinary citizens in the country do subscribe to the philosophy that developed nations should not dump their hazardous wastes in developing nations, the federal government has not yet become a formal member of the international organization espousing and acting on that view.

Environmental Justice and Climate Change

Individuals and groups interested in environmental justice on both a national and an international scale have recently turned their attention to a new facet of the problem: global climate change. Since the 1970s, evidence has been collecting to indicate that the world's average annual temperature is rising. Most experts agree that the primary reason for this change is the release of carbon dioxide and other so-called greenhouse gases produced by human activities that trap heat reflected from the Earth's surface. Increases in Earth's average annual temperature, most experts believe, are likely to result in a number of significant changes in the planet's surface characteristics, such as melting of glaciers, ice sheets, and other accumulations of ice, as well as a significant warming of the oceans. These changes, in turn, may produce dramatic effects on human civilizations, such as drowning of coastal cities and some low-lying islands; changes in precipitation and local temperature patterns; alterations in the distribution of natural plant and animal species; changes in disease patterns among humans, other animals, and plants; and new patterns of severe storm systems, such as hurricanes, typhoons, cyclones, and tornadoes.

In 2007, the Intergovernmental Panel on Climate Change (IPCC) released its most recent report on these changes (Intergovernmental Panel on Climate Change 2007a). It concluded, as in earlier reports, that some aspects of climate change are well documented and virtually beyond dispute (such as changes in global temperature patterns), that other data are generally accepted, although legitimate disputes still exist as to their significance (such as the causes of temperature change), and that still other conclusions about climate change are subject to considerable disagreement and dispute (such as possible effects on the human and natural environment of temperature changes).

Some of the most pessimistic commentators predict very troubling consequences of global climate change. A 2008 report by Cable News Network highlighted some of these predictions. The report warned that

- more than a billion people worldwide will face water shortages because of the melting ice pack;
- about a billion people will lose their livelihoods because of desertification of their lands;
- more than 200 million people will become refugees because of rising sea levels, erosion, and loss of agricultural lands;
- in one country alone, Bangladesh, about 17 million people could lose their homes because of flooding and severe storms;
- the number of Africans at risk for malaria will increase by more than 60 million if temperatures rise by 2°C; and
- more than 180 million residents of sub-Saharan Africa could die of diseases "directly attributable" to climate change. (Oliver 2008)

Nearly all aspects of the global climate change problem have strong racist and classist elements. In December 2006, Oxford University professor J. Timmons Roberts and Bradley C. Parks of the Millennium Challenge Corporation in Washington, D.C., published a thorough-going analysis of this problem in their book, *A Climate of Injustice: Global Inequality, North-South Politics, and Climate Policy* (Roberts and Parks 2006). They showed that, although the richest nations in the world produce more than 60 percent of all greenhouse gases released into the atmosphere, developing nations face by far the greatest risk of damages caused by climate change.

A number of factors contribute to the greater risk faced by poor people and people of color throughout the world as a result of climate change. One factor is simply geographical location. Many of the areas most likely to be affected by rising ocean levels are occupied by people of color: the low-lying regions of the Indian subcontinent and many small Pacific islands. Among the first communities of color to be threatened by rising ocean levels are likely to be those residents of small island nations, such as Kiribati, the Maldives, Nauru, Tuvalu, and Vanuatu. Some of these island nations are no more than a few meters in elevation at their highest points, and residents may have no choice other than to abandon their homelands to avoid being swamped by the ocean. In fact, the first population to have been relocated as a result of rising ocean waters was probably the residents of the village of Lateu, on Vanuatu, who were evicted from their homes in December 2005 because of the risk of rising tides (Environment News Service 2005).

Similarly, populations living in coastal regions in East Asia (almost exclusively communities of color) are at high risk from more frequent and severe storms, such as cyclones that strike the area every year. The devastating cyclone that hit the Irrawaddy River delta of Myanmar (formerly Burma) on May 9, 2008, underlines the problems faced by people who are already poor when they are the victims of a severe storm. When even their own national government refused aid from other nations, these people suffered death, injury, and disease not only from the direct effect of winds and water but also from the loss of access to food, clean water, clothing, shelter, and medicine, some of which were lacking in their lives even before the cyclone struck.

Some of the specific consequences resulting from climate change predicted by the IPCC in its 2007 report are as follows:

- Between 75 million and 250 million people will be exposed to increased water stress due to climate change by the year 2020.
- In some countries, yields from rain-dependent crops could be reduced by up to 50 percent.
- Climate change will intensify already severe pressure on populations in all parts of Asia as a result of urbanization, industrialization, and economic development.
- The populations of eastern, southern, and southeastern Asia will face the threat of endemic morbidity and mortality as a result of increased diarrheal diseases associated primarily with floods and droughts.
- Extensive tropical forests in central South America will gradually convert to savannahs, and semi-arid regions will convert to arid land as the result of significant decreases in annual rainfall and increases in temperature. (Intergovernmental Panel on Climate Change 2007b)

Over the last decade, a number of environmental justice groups have begun to stress the environmental justice aspects of the climate change problem, referring in many cases to a campaign for "climate justice." One of the most important of these efforts was a statement of general principles adopted in August 2002 by a group of organizations meeting during the 2002 Earth Summit conference of the UN Environment Programme in Johannesburg, South Africa. In all, 14 organizations—CorpWatch, US; Friends of the Earth International; Global Resistance; Greenpeace International; Groundwork, South Africa; the Indigenous Environmental Network; Indigenous Information Network, Kenya; the National Alliance of People's Movements of India; National Fishworkers Forum, India; Oilwatch Africa; OilWatch International; Southwest Network for Environmental and Economic Justice, US; Third World Network, Malaysia; and World Rainforest Movement, Uruguay—signed the agreement, now known as the Bali Principles of Climate Justice. The statement was developed using Environmental Justice Principles adopted at the People of Color Environmental Justice Leadership Summit held in Washington, D.C., in 1991 as a blueprint.

Among the principles enunciated in the statement are demands that:

- "governments are responsible for addressing climate change in a manner that is both democratically accountable to their people and in accordance with the principle of common but differentiated responsibilities";

- "communities, particularly affected communities[,] play a leading role in national and international processes to address climate change";
- "fossil fuel and extractive industries be held strictly liable for all past and current life-cycle impacts relating to the production of greenhouse gases and associated local pollutants";
- governments recognize "the rights of victims of climate change and associated injustices to receive full compensation, restoration, and reparation for loss of land, livelihood and other damages";
- "the right of all workers employed in extractive, fossil fuel and other greenhouse-gas producing industries to a safe and healthy work environment without being forced to choose between an unsafe livelihood based on unsustainable production and unemployment";
- "the right to self-determination of Indigenous Peoples, and their right to control their lands, including subsurface land, territories and resources and the right to the protection against any action or conduct that may result in the destruction or degradation of their territories and cultural way of life"; and
- "the right of indigenous peoples and local communities to participate effectively at every level of decision-making, including needs assessment, planning, implementation, enforcement and evaluation, the strict enforcement of principles of prior informed consent, and the right to say 'No.'" (India Resource Center 2002)

The analysis of climate change issues from an environmental justice perspective often highlights unexpected consequences of climate-control actions that otherwise receive favorable reviews—for example, one of the powerful tools developed as a result of the Framework Convention on Climate Change adopted at a conference held in Kyoto, Japan, in December 1997. According to this agreement, nations and corporations can reduce their measured contribution to climate change by engaging in carbon trades. A company that produces large amounts of greenhouse gases, for example, can have the overall contribution to climate change reduced by buying carbon credits from companies that pollute less (such as those, perhaps, found in LDCs) or by contributing to the development of climate change–ameliorating projects, such as the development of tree farms in undeveloped areas. But analysis often shows that such carbon-trading efforts actually result in a disproportionate effect on poor people of color. As one scholar has written, "Climate change policies addressing transportation, energy production, industry, commercial enterprises, housing, land use, and agriculture will inevitably have significant social and economic repercussions—on the poor, on consumers, and on affected industries. Notwithstanding the critical importance of significant greenhouse gas (GHG) reductions, policies designed in a vacuum, focusing solely on reductions, could create significant and unintentional adverse consequences. Moreover, policies to address climate change have the potential to address long-standing societal problems, like distributional inequities" (Kaswan 2008, 1; also see EJ Matters 2008; Smith 2007).

Concerns about the environmental justice aspect of climate change policies has resulted in the growth in the number of academic papers on the subject, activist efforts to influence domestic and international policy, the formation of new organizations, and other activities associated with the appearance of new political movements. For example, in the United States, a number of new organizations have appeared to study, educate, and lobby about issues in "climate justice." One of the most influential of these organizations is the Environmental Justice and Climate Change Initiative (EJCCI), located in Oakland, California. EJCCI is a consortium of 28 environmental, civil rights, and environmental justice groups, including the Black Leadership Forum, Communities for a Better Environment, CorpWatch, the Deep South Center for Environmental Justice at Xavier University, the Indigenous Environmental Network, Southern Organizing Committee, Southwest Network for Economic and Environmental Justice, and West Harlem Environmental Action.

In 2001, EJCCI developed a document titled "10 Principles for Just Climate Change Policies in the U.S." Those principles were as follows:

1. Stop Cooking the Planet
2. Protect and Empower Vulnerable Individuals and Communities
3. Ensure Just Transition for Workers and Communities

4. Require Community Participation
5. Global Problems Need Global Solutions
6. The U.S. Must Lead
7. Stop Exploration for Fossil Fuels
8. Monitor Domestic and International Carbon Markets
9. Caution in the Face of Uncertainty
10. Protect Future Generations (Environmental Justice and Climate Change Initiative 2008)

EJCCI says that these 10 principles form the guidelines for changing the focus of the debate over climate change in the United States, for attempting to improve the accountability of industry and the U.S. government in matters of climate change, for involving more youth in the campaign against climate change, and for driving its efforts for fund-raising.

As efforts to control human activities that may be responsible for climate change increase, efforts to monitor the disproportionate effects on low-income and minority communities around the world are almost certain to continue and, most likely, intensify.

Environmental Justice and Sustainable Development

Perhaps the most popular catchphrase making the rounds among environmentalists on an international scale today is *sustainable development*. The term first became generally known as the result of its use in the UN World Commission on Environment and Development's report *Our Common Future* (also known as the Brundtland Report), published in 1987. The report defined sustainable development as "development that meets the needs of the present without compromising the ability of future generations to meet their own needs" (World Commission on Environment and Development 1987, 43). The concept of a world in which all people live safely and comfortably without devastating the environment is an overwhelmingly attractive idea. However, the precise mechanisms by which such a goal can be reached have never been entirely clear. As the editor of a special issue of the *Duke Environmental Law & Policy Forum* has observed, "Sustainable development is a little like Zen. Everybody talks about it, but few people really know what it is" (Murray 1999, 148).

Over the past two decades, the significance of sustainable development as a guiding principle in policy making around the world has been increasingly accepted by international, regional, and national organizations and agencies. For example, the World Summit for Social Development held in Copenhagen in 1995 proclaimed that "[w]e are deeply convinced that economic development, social development and environmental protection are interdependent and mutually reinforcing components of sustainable development, which is the framework for our efforts to achieve a higher quality of life for all people" (World Summit for Social Development 2008).

In spite of any uncertainty about the exact meaning of the term *sustainable development*, its connection with the field of environmental justice has long been clear and unquestionable to many observers. The Copenhagen Declaration went on to connect sustainable development specifically with the principles of environmental justice: "Equitable social development that recognizes empowering the poor to utilize environmental resources sustainably is a necessary foundation for sustainable development. We also recognize that broad-based and sustained economic growth in the context of sustainable development is necessary to sustain social development and social justice" (World Summit for Social Development 2008).

The task for individuals and organizations, then, is to find the specific points at which the goals and methods of sustainable development intersect with those of environmental justice. A report by the Center for International Environmental Law (CIEL), *One Species, One Planet*, suggests three points of confluence:

- the right to life, including the right to a healthy environment;
- the traditional and customary property rights of indigenous and other local communities, especially those in the global south; and
- participatory and procedural rights. (Center for International Environmental Law 2002)

A significant portion of *One Species, One Planet* is devoted to an analysis of the way in which environmental justice and sustainable development intersect in these three areas. In the first case, the report emphasizes the fact that the right to life is the most basic of all human rights and, by its very nature, implies a right to a healthy environment. Without such an environment, a healthy life, and even life itself, may not be possible. Many nations have now adopted constitutional or legal guarantees for a safe and healthy environment that may also include the right to a safe workplace, the right to organize, the right to an adequate standard of living, and the right to participate in decision making that affects peoples' lives.

One Species, One Planet points out the importance of community rights in addition to the rights of individual persons. It takes special note of the fact that, in many nations, this concept is still largely lacking. Those nations may acknowledge the individual rights of its citizens but ignore the special and legitimate claims of ethnic, racial, linguistic, or other groups. Such a philosophy is especially injurious to communities dependent on natural resources for their survival. When a national government places resource recovery over the needs of individual communities, the rights of those communities have been violated. Any number of examples of this political philosophy can be found in many countries of the world, not the least of which is the United States. In this country, the federal government has long regarded lands owned by Native Americans as fair game in the search for and extraction of natural resources such as coal, oil, and uranium, even though such activities may profoundly disrupt the lives of Native Americans.

The CIEL report also reviews the concept of *community-based property rights* (CBPRs), which has been developed by legal experts interested in environmental justice and sustainable development. CBPRs are quite different from more traditional and familiar "common property" and "community-based natural resource management" legal strategies in that they place much more emphasis on the rights of an ethnic, racial, or other group. "The *distinguishing feature* of CBPRs," the report explains, "is that their exercise derives its authority from the community in which they exist, not from the state where they are located" (Center for International Environmental Law 2002, 14 [emphasis in original]).

Finally, the report explains the critical role of community participation in making decisions related to community survival and livelihood. Currently, hundreds of millions of natural resource–dependent poor and indigenous people, primarily in the global south, have essentially no control over the environmental factors that affect their lives. They are almost entirely at the mercy of national governments and, sometimes, transnational corporations that make decisions about resource allocation on the basis of factors that have little or no connection with the lives of local inhabitants. Neither environmental justice nor sustainable development can have any meaning at all unless local communities develop the skills for analyzing and solving the environmental and social problems they face. To achieve this goal, the CIEL report argues, national governments must accept the importance of and be willing to promote the concept of participatory democracy among their poor and indigenous communities. Implied by this term are a number of individual rights, such as the right to receive and disseminate information, the right to take part in planning and decision making, and the right of access to administrative and judicial justice.

One Species, One Planet is only one of many documents that have tried to establish a philosophical basis for the intersection of sustainable development and environmental justice. There appear to have been fewer instances in which national, regional, state, or local governments have tried to develop policies and programs that actually put this philosophy into practice. One interesting exception is a report released by the Scottish government in December 2005 titled *Choosing Our Future: Scotland's Sustainable Development Strategy*. The report attempts to show how the principles of sustainable development can be used to inform and drive culture, policy, and action in Scotland. Chapter 8 of the report deals specifically with the intersection between sustainable development and environmental justice. After explaining that a major emphasis of the report is to "improv[e] the quality of life of individuals and communities in Scotland, securing environmental justice for those who suffer the worst local environments," the report lists a number of specific actions through which those goals can be achieved. They include reducing the country's dependence on landfills (which have disproportionate effects on nearby communities); improving transport systems in such a way as to more evenly distribute the costs of road traffic, congestion, and air pollution; tackling problems of fuel poverty and unsafe construction in communities dealing with these problems; providing more parks and green spaces to make neighborhoods safer and healthier places in which to live and work; improving protection from floods; enhancing public access to information about their environment and to participation in decision making; and developing new educational programs to provide special benefits to communities where they are needed (Scottish Executive 2005, 42).

The challenge for both environmental justice and sustainable development groups today is to find specific ways in which objectives from both areas of concern can be achieved at the same time. People around the world are trying various programs to achieve this goal. As only one example, concerned citizens in the Dudley Street neighborhood of Roxbury in Boston have founded the Dudley Street Neighborhood Initiative (DSNI) to work for sustainable development of their community based on the principles of environmental justice. The organization was created by members of the community in 1984 "out of fear and anger to revive their neighborhood that was devastated by arson, disinvestment, neglect and redlining practices, and protect it from outside speculators" ("Dudley Street Neighborhood Initiative" 2008).

By the early 1980s, the Dudley Street neighborhood had become a classic run-down, crime-infested urban slum with a population that was about three-quarters African American and Cape Verdean, about one-quarter Latino, and less than 5 percent white. About 1,300 parcels of land, nearly a third of the district's 60 acres, had been abandoned by their owners. The vacant land had become a convenient location for trash companies and individuals to deposit their garbage, and the neighborhood soon became known as "Boston's dump yard" (Porter 2005).

In contrast to many urban neighborhoods of this kind, however, Dudley residents decided to do something about their increasingly desperate situation. The organization they formed has since become a model of what highly motivated activists can achieve when they use the best features of both the environmental justice and sustainable development movements. Within three years, DSNI had become the first (and so far only) community group in the United States to achieve the right to use the power of eminent domain over property within its area, had formed a land trust to administer that property, and had been the first inner-city neighborhood to win a state grant to create a town common. Over the succeeding two decades, it built on these early beginnings, eventually developing more than 600 of the original vacant lots and building more than 300 units of affordable housing in the process.

Typical of DSNI's commitment to the melding of environmental justice and sustainable development is the organization's urban agricultural project in which rehabilitated brownfields have been turned into urban gardens. In 2002, the group received a $25,000 grant from the EPA for its Food Project. The money went for training, educating, and employing youth in the neighborhood to practice sustainable agriculture and integrated pest management as ways of creating a sustainable food supply within the context of an environmental justice philosophy. DSNI also continues to work with 14 waste-related industries in the neighborhood to develop systems for waste reduction and waste recycling. For its achievements in combining the best features of environmental justice and sustainable development, DSNI has received a number of awards, including Boston's Best Kept Neighborhood Civic Award, the 1997 American Planning Association Award for housing planning, and the 2001 American Express Building Leadership Award. Without much doubt, as prestigious as these awards may be, the greatest accomplishment of DSNI has been the revitalization and enhancement of its own community. Today, DSNI remains a powerful force in the integration of sustainable development and environmental justice principles in the Boston metropolitan area.

Stories similar to that of the Dudley Street neighborhood are now beginning to appear in low-income and people-of-color communities around the world. They illustrate some of the great accomplishments that can be realized when the principles of environmental justice are melded with the ideals of sustainable development in enhancing the lives of people who have long faced profound social, health, political, and environmental challenges.

References

Alter, Harvey. 1995. "Halting the Trade in Recyclable Wastes Will Hurt Developing Countries." In *Environmental Justice,* edited by J. S. Petrikin, chap. 9. San Diego: Greenhaven Press.

Basel Action Network. 2008. "About the Basel Ban/A Chronology of the Basel Ban." [Online article or information; retrieved May 18, 2008.] http://www.ban.org/about_basel_ban/chronology.html.

Basel Convention on the Control of Transboundary Movements of Hazardous Wastes and Their Disposal. 2008. "Ban Amendment to the Basel Convention on the Control of Transboundary Movements of Hazardous Wastes and Their Disposal Geneva, 22 September 1995." [Online article or information; retrieved May 20, 2008.] http://www.basel.int/ratif/ban-alpha.htm.

Center for International Environmental Law (CIEL). 2001 (August 9). "Green Groups Call on USA to Ratify International Toxic Waste Dumping Ban as Part of Basel Treaty." http://www.ciel.org/Chemicals/Ban_BASEL_9Aug01.html.

Center for International Environmental Law (CIEL). 2002. *One Species, One Planet: Environmental Justice and Sustainable Development.* Washington, DC: Center for International Environmental Law.

Clapp, Jennifer. 2001. *Toxic Exports: The Transfer of Hazardous Wastes from Rich to Poor Countries.* Ithaca, NY: Cornell University Press.

Cook, Jaclyn, and Eddie Koch, eds. 1992. *Going Green: People, Politics and the Environment in South Africa.* Cape Town, South Africa: Oxford University Press.

"Dudley Street Neighborhood Initiative." 2008. [Online article or information; retrieved May 24, 2008.] http://www.dsni.org/.

EJ Matters. 2008. "The California Environmental Justice Movement's Declaration on Use of Carbon Trading Schemes to Address Climate Change." [Online article or information; retrieved May 20, 2008.] http://www.ejmatters.org/declaration.html.

Environment News Service. 2005 (December 6). "Pacific Island Villagers Become Climate Change Refugees." [Online article or information; retrieved 5/20/08.] http://www.ens-newswire.com/ens/dec2005/2005-12 –06–02.asp.

Environmental Justice and Climate Change Initiative. 2008. "10 Principles for Just Climate Change Policies in the U.S." [Online article or information; retrieved May 20, 2008.] http://www.ejcc.org/10%20Principles%20of%20Climate%20Justice.pdf.

European Economic Community. 2008. "Council Regulation (EEC) No 259/93 of 1 February 1993 on the Supervision and Control of Shipments of Waste within, into and out of the European Community." [Online article or information; retrieved May 20, 2008.] http://eur-lex.europa.eu/LexUriServ/LexUriServ.do?uri=CELEX:31993R0259:EN:NOT.

Fondaw, Corey. 2008. "Environmental Justice Case Study: Thor Chemicals and Mercury Exposure in Cato-Ridge, South Africa." [Online article or information; retrieved May 18, 2008.] http://www.umich.edu/~snre492/Jones/thorchem.htm.

Friends of the Earth. "The Bakun Hydroelectric Project: Impacts on the Indigenous People." [Online article or information; retrieved 5/20/08.]

http://www.foe.co.uk/resource/briefings/bakun_indigenous_people
.html. Last modified November 1996.

Greenpeace. 2008. "Philadelphia Ash Dumping Chronology." [Online
article or information; retrieved May 18, 2008.] http://www.essential
action.org/return/chron.html.

India Resource Center. 2002 (August 28). "Bali Principles of Climate
Justice." http://www.indiaresource.org/issues/energycc/2003/bali
principles.html.

Intergovernmental Panel on Climate Change. 2007a. *Climate Change
2007: Synthesis Report.* Geneva: Intergovernmental Panel on Climate
Change. [Online article or information; retrieved May 20, 2008.]
http://www.ipcc.ch/pdf/assessment-report/ar4/syr/ar4_syr.pdf.

Intergovernmental Panel on Climate Change. 2007b. *Climate Change
2007: Synthesis Report: Summary Report.* Geneva, Switzerland: Intergovernmental Panel on Climate Change. [Online article or information; retrieved May 20, 2008.] http://www.ipcc.ch/pdf/assessment-report/
ar4/syr/ar4_syr_spm.pdf.

Johnston, Barbara Rose. 1994. *Who Pays the Price: The Sociocultural Context of Environmental Crises.* Washington, DC: Island Press.

Kaswan, Alice. 2008. "Environmental Justice and Domestic Climate
Change Policy." *Environmental Law Reporter* (May). Posted online January
18. [Online article or information; retrieved 5/20/08] http://papers.ssrn
.com/sol3/papers.cfm?abstract_id=1077675.

Knight, Danielle. 2001 (March 29). "Foreign Mercury Waste Still Contam-
inating South Africa." [Online article or information; retrieved 5/20/08.]
http://www.afrol.com/News2001/sa017_mercury_waste.htm.

Maasai Environmental Resource Coalition (MERC). 2008. [Online article
or information; retrieved May 16, 2008] http://www.maasaierc.org/
missionandprojects.html.

Malaysian Timber Council (MTC). "Bakun Hydroelectric Project." [On-
line article or information; retrieved 5/20/08.] http://www.mtc.com
.my/publication/library/bakun/bakun1.htm. Last modified on January 25, 2002.

Marshall, Will. 2002 (April 9). "Australian Mining Giant Leaves Environmental Disaster in Papua New Guinea." World Socialist Web Site. [On-
line article or information; retrieved 5/20/08.] http://www.wsws.org/
articles/2002/apr2002/png-a09.shtml.

Mpanya, Mutombo. 1992. "The Dumping of Toxic Waste in African
Countries: A Case of Poverty and Racism." In *Race and the Incidence of
Environmental Hazards: A Time for Discourse,* edited by Bunyan Bryant
and Paul Mohai, chap. 15. Boulder, CO: Westview Press.

Murray, Kali N., ed. 1999. "Introduction." *Duke Environmental Law & Policy Forum* 9 (Spring): 147–152.

Narimatsu, Julie. 2008. "Environmental Justice Case Study: Maasai Land Rights in Kenya and Tanzania." [Online article or information; retrieved May 16, 2008.] http://www.umich.edu/~snre492/Jones/maasai.htm.

Norris, Ruth, ed. 1982. *Pills, Pesticides & Profits: The International Trade in Toxic Substances.* Croton-on-Hudson, NY: North River Press.

Oliver, Rachel. 2008 (February 18). "Rich, Poor and Climate Change." CNN.com. [Online article or information; retrieved 5/20/08.] http://edition.cnn.com/2008/BUSINESS/02/17/eco.class/.

Porter, Tim. 2005 (September 22). "One Neighborhood, Many Stories." Poynternonline. http://www.poynter.org/column.asp?id=69.

Roberts, J. Timmons, and Bradley C. Parks. 2006. *A Climate of Injustice: Global Inequality, North-South Politics, and Climate Policy.* Cambridge, MA: MIT Press.

Schwartz, Jerry. 2000 (September 5). "The Full Story of the *Khian Sea* and the Gonaives Ash Mountain." [Online article or information; retrieved 5/20/08.] http://www.webster.edu/~corbetre/haiti-archive/msg05049.html.

Scottish Executive. 2005. *Choosing Our Future: Scotland's Sustainable Development Strategy.* Edinburgh, UK: Scottish Executive.

Smith, Kevin. 2007. *The Carbon Neutral Myth: Offset Indulgences for Your Climate Sins.* Amsterdam: Transnational Institute. [Online article or information; retrieved 5/20/08.] http://www.carbontradewatch.org/pubs/carbon_neutral_myth.pdf.

Steger, Tamara. 2007. *Making the Case for Environmental Justice in Central & Eastern Europe.* Budapest, Hungary: CEU Center for Environmental Policy and Law, the Health and Environment Alliance, and the Coalition for Environmental Justice.

Wade, Robert. 1997. "Greening the Bank: The Struggle over the Environment (1970–1995)." In *The World Bank: Its First Half Century,* edited by Devesh Kapur, John P. Lewis, and Richard Webb, chap. 13. Washington, DC: The Brookings Institution.

Whirled Bank Group. 2008. "Larry Summers: The Bank Memo." [Online article or information; retrieved May 18, 2008.] http://www.whirledbank.org/ourwords/summers.html.

World Commission on Environment and Development. 1987. *Our Common Future.* Oxford, UK: Oxford University Press.

World Summit for Social Development. 2008. "Copenhagen Declaration on Social Development." [Online article or information; retrieved May 20, 2008.] http://www.un-documents.net/cope-dec.htm.

4

Chronology

T he environmental justice movement is relatively young. Its origin is often traced to the protests held in Warren County, North Carolina, in 1982. Yet, many of the elements of the modern movement can be traced back much further, at least to the Civil Rights Act of 1875 and the first environmental law of 1899. This chapter summarizes some of the most important events in the history of the modern environmental justice movement.

1899   America's first environmental law, the Refuse Act of 1899, is passed. The act is originally written to protect navigable waters from pollution by sediments, disease-carrying organisms, and oil discharges, but it is broad enough to cover many environmental problems through the early 1970s.

1947   The Federal Insecticide, Fungicide, and Rodenticide Act of 1947 is passed by Congress. The act originally deals only with pesticides shipped across state lines. The act is amended and extended in 1972 to cover broader aspects of pesticide production and use.

1948   The Water Pollution Control Act of 1948 is passed by Congress. The statute provides a mechanism by which the federal government can work with state and local governments to reduce or eliminate pollution of interstate waterways and to improve the

1948 (cont.)	quality of surface and underground water resources. The act is never implemented because almost no funds are appropriated for it.
1956	A stronger version of the 1948 Water Pollution Control Act is adopted. The act is amended and strengthened a number of times in following years.
1960	The U.S. Supreme Court rules that segregation in interstate bus and train stations is illegal. A year later, the Interstate Commerce Commission officially prohibits such practices.
1964	The 24th Amendment to the U.S. Constitution is adopted. The amendment abolishes the poll tax, a primary means by which people of color were long prevented from voting.
	The Civil Rights Act of 1964 is adopted, guaranteeing to all American citizens equal access to employment and public accommodations, regardless of race, creed, color, sex, or national origin.
1965	The Voting Rights Act of 1965, strengthening its predecessor, the Voting Rights Act of 1957, is passed.
1968	The Fair Housing Act of 1968 is adopted, prohibiting discrimination based on race, color, religion, or national origin in rental or purchase of homes, real estate advertising, mortgage lending, homeowner's insurance, and zoning.
1969	Ralph Abascal, a young attorney and general counsel to California Rural Legal Assistance, files suit on behalf of six migrant farmworkers that eventually results in a ban on the use of the pesticide DDT.
1970	Congress passes and President Richard M. Nixon signs the National Environmental Policy Act (NEPA) of 1969. The act is the single most comprehensive legislative action dealing with environmental issues in the nation's history. It establishes, for the first time, a

comprehensive environmental policy for the U.S. government and all of its constituent agencies. It also creates the President's Council on Environmental Quality, the primary responsibility of which is to develop regulations that will implement the law's policies. Of perhaps greatest significance, NEPA requires that all federal agencies submit environmental impact statements that describe the impact of major agency actions on the human, biological, and physical environment.

Congress passes and the president signs the Clean Air Act of 1970, the strongest and most comprehensive air pollution legislation to date in the United States. The two most important features of the law are the creation of new standards for the presence of air pollutants—the New Source Performance Standards (NSPS) and National Emissions Standards for Hazardous Air Pollutants (NESHAP). These standards express maximum tolerable limits that will protect human health for a variety of pollutants.

The U.S. Environmental Protection Agency (EPA) is created as an independent agency in the Reorganization Plan No. 3 of 1970.

Sen. Philip Hart (D-MI) arranges a meeting among environmental groups, labor unions, and minority organizations to consider issues of the urban environment of concern to all such groups.

April 22 is declared to be Earth Day. By some estimates, more than 20 million people demonstrate on behalf of improved environmental conditions in the United States and around the world.

1971   The annual report of the Council on Environmental Quality acknowledges that environmental quality is affected by the racial and class status of communities.

1972   The Water Pollution Control Act of 1972 is adopted. The act represents a giant step forward in the nation's

1972
(cont.)

commitment to protect its water resources from pollution. Its primary achievement is to consolidate and reorganize provisions for dealing with water pollution contained in 11 preceding laws dealing with various aspects of water quality.

1975

The Federal Insecticide, Fungicide, and Rodenticide Act Amendments of 1975 are adopted. These amendments update the Federal Insecticide, Fungicide, and Rodenticide Act of 1947, as amended in 1964, and the Federal Environmental Pesticide Control Act of 1972. These laws are further updated and extended in 1980, 1988, 1990, 1991, and 1996. In each case, the authority of the federal government to regulate and control the production, sale, and use of a variety of pesticides is made more comprehensive.

1976

The Toxic Substances Control Act (TSCA) of 1976 and Resource Conservation and Recovery Act (RCRA) of 1976 are adopted. TSCA is the first comprehensive legislation giving the federal government authority to regulate new and existing chemical substances. RCRA requires federal agencies to develop methods for assessing the impact of hazardous wastes on the physical, biological, and human environment.

A conference on the urban environment is held at Black Lakes, Michigan. The conference is organized by the United Auto Workers and includes representatives from unions, environmental organizations, religious groups, and those concerned with economic justice.

The Urban League creates Project Que: Environmental Concerns in the Inner City.

1977

The Clean Air Act and Clean Water Act are revised and updated.

Grants totaling $66,000 permit the Urban Environment Conference to fund 11 nationwide conferences on issues related to environmental justice.

1978   Lois Gibbs first takes action on the health threats
           posed by hazardous wastes buried beneath the Nia-
           gara Falls, New York, neighborhood in which she
           lives. Gibbs's efforts eventually lead to the abandon-
           ment of the Love Canal area in which she lived and a
           cleanup that eventually leads to the relocation of
           more than 800 families.

1979   "City Care: A Conference on the Urban Environ-
           ment" is held in Detroit. Organizers include the Na-
           tional Urban League, the Urban Environment
           Conference, and the Sierra Club. One result of the
           meeting is a sense that traditional environmental
           groups are going to be uncomfortable with the inclu-
           sion of social justice issues in their agendas.

           Residents of the Northwood Manor neighborhood of
           Houston sue Southwestern Waste Management, Inc.,
           to prevent the construction of a waste disposal site in
           their community. The 82 percent black neighborhood
           claims that it is being unfairly discriminated against
           by the company's plans. The community eventually
           loses the lawsuit, but *Bean v. Southwestern* becomes a
           landmark case in that it is the first time a claim of en-
           vironmental discrimination forms the basis of legal
           action.

1982   Residents of Warren County, North Carolina, sup-
           ported by the United Church of Christ, stage a
           demonstration in opposition to the siting of a poly-
           chlorinated biphenyl (PCB) landfill near the commu-
           nity of Afton. More than 500 African American
           protestors are arrested in an unsuccessful attempt to
           block construction of the landfill. This event is
           widely regarded as the real beginning of the modern
           environmental justice movement.

1983   "Taking Back Our Health—An Institute on Surviving
           the Toxic Threat to Minority Communities" is held in
           New Orleans. The conference is sponsored by the Ur-
           ban Environment Conference shortly before it loses
           its funding and goes out of business.

1983
(cont.)

A report by the General Accounting Office (now the Government Accountability Office; GAO) states that three out of four hazardous waste sites in EPA Region 4 (Alabama, Florida, Georgia, Kentucky, Mississippi, North Carolina, South Carolina, and Tennessee) are located in African American communities.

The Olin Corporation announces that it will fund the cleanup of a waste site near Triana, Alabama, that had been contaminated by DDT wastes from the Redstone Arsenal. The action is the first time that an EPA order has resulted in a health care settlement for a specific local population. The total cost of the operation is about $25 million. Triana had earlier been called the "unhealthiest town in the United States" as a result of the contaminated site.

1985

The grassroots committee, People Concerned about MIC (methyl isocyanate), is organized in Institute, West Virginia. Consisting largely of African Americans, the group is created in response to a chemical leak from a nearby Union Carbide plant that resulted in about 135 residents being sent to the hospital.

The EPA commissions the Council of Energy Resource Tribes to conduct a study of potential hazardous waste sites located on or near Indian lands. The survey reveals as many as 1,200 hazardous waste sites on or near 25 Indian reservations.

1986

Activists in the West Harlem neighborhood of New York City organize to take action against pollution being produced by the city's North River Sewage Treatment plant. Two years later, the organization is formalized as West Harlem Environmental Action (WE ACT), still one of the strongest and most active neighborhood environmental justice groups in the United States.

Juana Beatriz Gutierrez leads a group of her neighbors in protesting the construction of a new state prison in their community. They are eventually successful, and the state decides in 1992 not to build the structure. The protestors eventually organize as the Las Madres del Este de Los Angeles (Mothers of East Los Angeles).

1987   The Commission for Racial Justice of the United Church of Christ publishes a report, "Toxic Wastes and Race in the United States," showing that race, even more than income level, is the critical factor shared by communities exposed to toxic wastes.

Robert D. Bullard publishes *Invisible Houston*, one of the first books to describe in detail environmental racism and social injustice in black neighborhoods of Houston.

Italian businessmen Gianfranco Raffaeli and Renato Pent sign an illegal agreement with Nigerian businessman Sunday Nana, to use his property for the disposal of hazardous wastes from Italy. The wastes are shipped under the guise of building materials and waste chemicals. More than two dozen workers become ill as a result of handling the wastes in Nigeria.

The government of Haiti issues a permit for the importation of fertilizer from the United States onboard the freighter *Khian Sea*. The shipment contains, in fact, not fertilizer, but 13,476 tons of toxic wastes from a Philadelphia incinerator.

1988   The U.S. Supreme Court rules that the development of geothermal energy plants in the Hawaiian Islands does not impinge on the First Amendment right to freedom of religion among native Hawaiians. Native Hawaiians had claimed that the development of such plants would destroy the rainforests and interfere with their worship of the volcanic goddess Pele.

A shipment of garbage and incinerator ash from Philadelphia, originally rejected by both Haiti and Panama aboard the *Khian Sea*, is accepted by the Guinean government. Later reports claim that the

1988 *(cont.)*	trees on Guinea's Kasa Island, where the shipment is dumped, turn brown and die.
1989	Residents of the "Cancer Alley" section of the lower Mississippi River organize the Great Louisiana Toxic March in an attempt to bring attention to their polluted living conditions.

Representatives of a number of Indian tribes meet to discuss issues related to natural resource extraction and environmental issues affecting Native Americans. This and succeeding meetings lead to the creation of the Indigenous Environmental Network in 1990.

Las Madres del Este de Los Angeles successfully protest the construction in their neighborhood of a $29 million incinerator designed to burn 125,000 pounds of toxic wastes per day.

Thirty-three nations sign the Basel Convention on the Control of Transboundary Movements of Hazardous Wastes and Their Disposal (the "Basel treaty"). To date, 170 nations have signed the treaty, and 63 have ratified it.

1990   Students at the Harvard Law School, the New York University Law School, the University of California at Berkeley Law School, and Washington University in St. Louis hold conferences on the issue of environmental justice.

The first national gathering on environmental justice is held at the University of Michigan in Ann Arbor. The attendees at that conference, the so-called Michigan Coalition, issue a report titled "Race and the Incidence of Environmental Health." They send a letter to William Reilly, administrator of the EPA, "demanding action on environmental risks in minority and low-income communities and on tribal lands." In response to the letter, Reilly appoints the Environmental Equity Workgroup.

A number of grassroots environmental justice organizations, including the Gulf Coast Tenants Organization, the Southwest Organizing Project, the United Church of Christ Commission for Racial Justice, and the Southern Organizing Committee for Economic and Social Justice, jointly write to the so-called "big 10" of mainstream environmental groups, expressing their concerns about the groups' historic lack of attention to the special needs and interests of people of color and other minorities and challenging them to end their racist and elitist policies and practices.

New York City adopts a "fair share" act intended to ensure that every part of the city receives a fair share of hazardous facility sitings.

Robert D. Bullard publishes one of the classic texts on environmental justice, *Dumping in Dixie.*

1991   The Dow Chemical company pays about $10 million to relocate 110 homes near its plant in Morrisonville, Louisiana. The company decides it is cheaper to move residents than to invest in pollution abatement equipment at the plant.

The First National People of Color Environmental Leadership Summit is held in Washington, D.C. Attendees adopt the 17-point "Principles of Environmental Justice," which has served as a guideline for the movement ever since.

In what is apparently the first significant contribution from the law community on the issue of environmental justice, Rachel D. Godsil, a student, publishes a commentary on the subject in the *University of Michigan Law Review.* The article is titled "Remedying Environmental Racism."

The U.S. Agency for Toxic Substances and Disease Registry (ATSDR) initiates a number of actions relating to environmental inequities, including the

1991
(cont.)
Minority Environmental Health Conference and a study of minority communities located near National Priorities List hazardous waste landfills.

1992
The Charles Stewart Mott Foundation provides a grant for the publication of the first edition of the *People of Color Environmental Groups Directory*. It listed more than 300 such groups.

Students at Columbia University and the Universities of Michigan and Minnesota hold conferences on issues of environmental justice.

The EPA establishes the Office of Environmental Justice.

The EPA releases a report, "Environmental Equity: Reducing Risks for All Communities."

A report in the *National Law Journal*, "Unequal Protection: The Racial Divide in Environmental Law," claims that the EPA pursues discriminatory practices in enforcement of environmental laws and regulations.

A national workshop, "Equity in Environmental Health: Research Issues and Needs," is held at Research Triangle Park, North Carolina, under the sponsorship of the EPA, ATSDR, and the National Institute for Environmental Health Sciences.

Rep. John Lewis (D-GA) and Sen. Al Gore Jr. (D-TN) introduce the Environmental Justice Act of 1992.

Rep. Cardiss Collins (D-IL) introduces an amendment to the Resource Conservation and Recovery Act that would require a "community information statement" for the construction of any new hazardous facility. The statement would include a description of the demography of the proposed site as well as a projected estimate of the impact of the facility on the area.

1993   On Earth Day, President Bill Clinton pledges that he will issue an executive order instructing federal agencies to take cognizance of the issues raised by the environmental justice movement and to take actions that will reduce the problems emphasized by that movement.

The EPA establishes the National Environmental Justice Advisory Council.

Representative Lewis and Sen. Max Baucus (D-MT) reintroduce the Environmental Justice Act.

Hearings on the proposed promotion of the EPA to cabinet status include discussion of environmental justice issues.

The Texas Air Control Board and the Texas Water Commission jointly create the statewide Task Force on Environmental Equity and Justice to deal with basic issues, such as the reasons hazardous facilities tend to be located in minority communities, policies and procedures of the two agencies that relate to issues of environmental inequities, methods by which the agencies can become more "user friendly" to communities of color, and data-gathering methods by which the government might become more aware of environmental inequities in hazardous facilities sitings.

Citizens of Wallace, Louisiana, in the state's notorious Cancer Alley, successfully defeat plans for the construction of a new rayon and pulp-processing plant in their community by Louisiana Formosa Plastic. Tensas Parish had rezoned part of the town to permit construction of the plant, but residents, concerned about health effects of the plant's waste products, were able to deter Formosa's plans to build the plant.

The Asian Pacific Environmental Network (APEN) is established to provide a mechanism by which Asian Americans and Pacific Islanders can express their

1993 *(cont.)*	concerns about environmental injustices suffered by their communities.

Nine farmworker organizations from the United States and the Caribbean with a combined membership of about 50,000 join to create the Farmworker Network for Economic and Environmental Justice (FNEEJ).

Environmental justice groups file the first law-suits under Title VI of the Civil Rights Act of 1964 against the Louisiana and Mississippi Departments of Environmental Quality. The suits are based on the premise that the two state agencies discriminate against low-income and minority communities in making environmental policies and decisions, thereby violating the provisions of Title VI. Over the next decades, more than a hundred such suits are filed against state, regional, and local agencies, although very few are successfully prosecuted.

1994 President Bill Clinton signs Executive Order 12898 on Environmental Justice ordering federal agencies to abolish and prevent policies that lead to a disproportionate distribution of environmental hazards to communities of color or low income.

The Interagency Symposium on Health Research and Needs to Ensure Environmental Justice, sponsored by the EPA, is held in Arlington, Virginia.

The Federal Interagency Working Group on Environmental Justice is created.

The Environmental Justice Resource Center (EJRC) at Clark Atlanta University, in Atlanta, is created with Robert D. Bullard as its director.

EJRC publishes the second edition of its *People of Color Environmental Groups Directory,* listing more than 600 environmental justice groups in the United States, Puerto Rico, Canada, and Mexico.

The United Church of Christ issues an update of its 1987 report, "Toxic Waste and Race Revisited," providing further evidence on the relationship between race and toxic waste facilities.

The Environmental Law and Justice Center at Texas Southern University's Thurgood Marshall School of Law is formed.

Researchers at the University of Massachusetts publish a study indicating that minorities and low-income people may not be disproportionately targeted in the construction of waste sites and that, in fact, such sites may actually have benefits for such communities. The study is funded by Waste Management, Inc., the largest solid waste disposal company in the world.

1995   The first public meeting of the Interagency Working Group on Environmental Justice is held at Clark Atlanta University.

The EPA announces the award of $3 million to assist 174 community-based organizations, tribal governments, and academic institutions to address environmental justice issues in their communities. The awards reflect an increase from the $500,000 awarded to 61 recipients in 1994.

A conference, "Environmental Justice and Transportation: Building Model Partnerships," is held in Atlanta, under the joint sponsorship of the U.S. Department of Transportation and the Clark Atlanta Environmental Justice Resource Center. The conference is a part of the department's public outreach plan as mandated by President Clinton's Executive Order 12898.

Six environmental justice networks—the Asian Pacific Environmental Network, Southwest Network for Environmental and Economic Justice, Indigenous Environmental Network, Farmworker Network for

1995
(cont.)

Economic and Environmental Justice, Southern Organizing Committee for Economic and Social Justice, and Northeast Environmental Justice Network—join to establish the Environmental Justice Fund, an alternative mechanism for funding grassroots environmental justice organizations.

Diné CARE, a Navajo Nation environmental justice group, prevails on the Bureau of Indian Affairs of the U.S. Department of the Interior to develop an environmental impact statement (EIS) and a 10-year forest management plan (FMP) for the Navajo Nation. Previously, both plans had been developed by the Peabody Coal Company, which was responsible for extensive mining projects on the tribal reservation.

The Asian Pacific Environmental network initiates its Laotian Organizing Project (LOP) to provide a means by which members of the Laotian immigrant community in Richmond, California, can express their views on environmental justice issues.

1996

The Seventh Generation Fund sponsors the Anti-Nuclear Summit in Albuquerque. Indigenous peoples from North America and the Pacific whose lives are affected by the production of nuclear materials agree on a joint declaration of their approach to nuclear issues.

Grassroots leaders in Pensacola, Florida, convince the EPA to relocate an entire community of 358 African American and/or low-income households living adjacent to the Escambia Wood Treatment waste treatment facility in Pensacola.

The African American Environmental Justice Action Network is formed as a collaborative activity of about 75 community groups working for environmental justice in 11 southern states.

The U.S. Institute of Medicine conducts a "toxic tour" of the so-called Cancer Alley in Louisiana as part of its data-gathering activities in preparation for its 1999 report *Toward Environmental Justice.*

1997   The People of Color and Disenfranchised Communities Environmental Health Summit is held in Waveland, Mississippi, to develop plans for dealing with the disproportionate effect of nuclear materials development activities on minority communities. As a result of the conference, the Department of Energy and other federal agencies begin to work with the coalition to build more positive relationships with such communities.

The "Healthy and Sustainable Communities Conference: Building Model Partnerships for the 21st Century" is held in Atlanta. The conference focuses on success stories in the environmental justice movement; conducts a review of best practices and model community initiatives; discusses successful tools, methodologies, and strategies; and provides opportunities for workers in the field to network with each other about common areas of interest.

The ATSDR of the Department of Health and Human Services (HHS) creates the Community/Tribal Subcommittee of the ATSDR Board of Scientific Counselors to advise it on issues relating to tribal communities that are affected by hazardous waste sites or other facilities.

Citizens against Nuclear Trash (CANT), an organization of grassroots activists from Forest Grove (population 150) and Center Springs (population 100), Louisiana, are successful in their efforts to prevent British Nuclear Fuels from constructing a uranium-processing plant in their neighborhood.

1998   The EPA issues its "Interim Guidance for Investigating Title VI Administrative Complaints Challenging Permits." The purpose of the document is to update the EPA's procedural and policy framework to deal with the increasing number of Title VI complaints

1998
(cont.)
that allege discrimination in the environmental permitting context. The guidelines meet with considerable opposition, and the EPA begins to revise them almost immediately.

The EPA rules on the first two Title VI complaints brought to its attention. In one, involving Shintech, Inc., the case becomes moot when the company decides to build its new polyvinyl plant at a different location. In the second, involving Select Steel, the EPA rules against the complainants. Select Steel is an important case because it is the first decision the EPA makes on Title VI claims.

Florida A&M University establishes the Environmental Justice and Equity Institute in Tallahassee.

The Wisconsin state legislature passes the Wisconsin Mining Moratorium Law, placing strict conditions on the issuance of permits for new metal sulfide mines. The law is strongly supported by Indian tribes because of their concerns about the adverse effects of mining on their lands and people.

1999
More than 300 black grassroots environmental justice activists from 37 states meet in New Orleans to discuss issues of common interest. They agree to form a new organization, which they call the Interim National Black Environmental and Economic Justice Coordinating Committee (INBEEJCC). In 2001, the organization changes its name to the National Black Environmental Justice Network (NBEJN).

The Institute of Medicine releases its landmark study on environmental inequities in the United States, *Toward Environmental Justice: Research, Education, and Health Policy Needs*.

Rep. James Clyburn (D-SC), chair of the Congressional Black Caucus, convenes a conference titled "Environmental Justice: Strengthening the Bridge between Economic Development and Sustainable Communities," held at Hilton Head, South Carolina. The conference is hosted by the Medical University of South Carolina's Environmental Biosciences Program.

California state senator Hilda L. Solis introduces the first state legislation on environmental justice to be adopted in the United States. Since that time, the state has developed an extensive and aggressive program for discovering and combating environmental inequities in California.

2000   The Environmental Justice Resource Center publishes the third edition of the *People of Color Environmental Groups Directory*, listing more than 1,000 environmental justice groups in the United States, Puerto Rico, Canada, and Mexico.

The Indigenous Environmental Network and Project Underground join forces to establish the Indigenous Mining Campaign Project, an effort to train indigenous people in dealing with mining issues on tribal lands.

The National Environmental Policy Commission (NEPC) is created, largely as a result of recommendations made by the Congressional Black Caucus (CBC) Environmental Justice Braintrust, led by Representative Clyburn (D-SC). The purpose of the NEPC is to develop an alternative environmental policy that takes into account environmental justice, human health, economic considerations, and other important factors.

The North Carolina General Assembly allocates $7 million to begin cleanup of the PCB landfill in Warren County, the environmental inequity issue that marked the beginning of the environmental justice movement in 1982.

2001   Greenpeace sponsors the "Celebrity Tour of Cancer Alley," which features actor Mike Farrell, Rep. Maxine Waters (D-CA), and writer Alice Walker. The tour

2001
(cont.)

revisits a number of infamous sites on the lower Mississippi River that have been denuded and devastated by industrial pollution and hazardous wastes.

A New Jersey district court rules in *South Camden Citizens in Action v. NJ Dept of Environmental Protection* that compliance with environmental laws is not equivalent to compliance with civil rights laws. The court rules that the state of New Jersey has violated Title VI of the Civil Rights Act of 1964. The case is significant because it is the first instance in which a complaint of this kind has prevailed. The district court's decision is later overturned by the Third Circuit of Appeals.

After nearly six years of litigation, the Monsanto Company agrees to a $42.8 million settlement with an environmental justice group called the Sweet Valley/Cobb Town Environmental Task Force, representing residents of a neighborhood of primarily black and low-income people adjacent to Monsanto's Solutia PCB plant. The production of PCBs was outlawed in the United States in 1978, and the Monsanto plant was only one of two facilities in the country where the chemicals were being made.

2002

After more than four decades of protests, negotiations, and litigation, the Shell chemical company agrees to pay for the relocation of more than 100 homeowners in the town of Norco, Louisiana. Some homes had been located less than 15 feet from the company's plant, and residents blamed the high rate of cancer and other disorders on exposure to emissions from the facility.

A total of 2,500 tons of toxic incinerator ash are buried at the Mountain View Reclamation Landfill, in Franklin County, Pennsylvania. The ash was part of a shipment of municipal wastes originally shipped from Pennsylvania on the freighter *Khian Sea* in 1986 for disposal in the Bahamas. When the Bahamian government, the Dominican Republic, Honduras,

Panama, Bermuda, Guinea Bissau, and the Dutch Antilles all refused to accept the wastes, the *Khian Sea* dumped 4,000 tons of the wastes on a beach near Gonaives, Haiti. It proclaimed that the wastes were fertilizer. Now, 14 years later, Waste Management, Inc., recovers the 2,500 tons of ash and returns it to the United States, where it is eventually buried in Pennsylvania.

The Second People of Color Environmental Leadership Summit is held in Washington, D.C. More than 1,400 activists attend the conference.

A coalition of environmental justice groups consisting of the Black Leadership Forum, the Southern Organizing Committee for Economic and Social Justice, the Georgia Coalition for the Peoples' Agenda, and Clear the Air release a report, *Air of Injustice: African Americans and Power Plant Pollution,* on the differential effects on African American communities of air pollution from power generation.

The University of Michigan School of Natural Resources and Environment sponsors a conference titled "Latinos and the Environment."

2003 The U.S. Commission on Civil Rights issues its report, *Not in My Backyard: Executive Order 12898 and Title VI as Tools for Achieving Environmental Justice.* The report attempts to assess the effectiveness with which four federal agencies—the EPA, the U.S. Department of the Interior, the U.S. Department of Housing and Urban Development, and the U.S. Department of Transportation—have implemented Executive Order 12898 and Title VI of the 1964 Civil Rights Act. They find that the four agencies have a relatively poor record in this regard.

The state of North Carolina completes its cleanup of the PCB toxic waste site in Warren County, where the environmental justice movement began more than two decades earlier.

2003
(cont.)

The New York State Department of Environmental Conservation issues a new policy that environmental justice concerns are to be included in decisions on the issuance of permits.

The University of Michigan School of Natural Resources and Environment establishes the Minority Environmental Leadership Development Initiative (MELDI) to enhance the leadership and career development opportunities available to minority students and minority environmental professionals.

The University of Michigan School of Natural Resources and Environment sponsors the conference "Environmental Justice and Climate Change."

2004

The U.S. Nuclear Regulatory Commission (NRC) publishes its policy on the treatment of environmental justice issues in the issuance of licenses for nuclear power plants. Critics suggest that the NRC policies will actually limit the attention it pays to environmental justice issues in general and to Executive Order 12898 in particular.

The Special Committee on Environmental Justice of the American Bar Association (ABA) publishes *Environmental Justice for All: A Fifty-State Survey of Legislation, Policies, and Initiatives,* summarizing statutes, policies, and other legal rulings on environmental justice. The report is also made available on the ABA's Web site, where it is updated on a regular basis.

The Office of Inspector General of the EPA issues its report, *EPA Needs to Conduct Environmental Justice Reviews of Its Programs, Policies, and Activities,* summarizing the agency's progress (or lack of it) in dealing with issues of environmental inequity.

The District Court for the Eastern District of Texas finds "a reasonable trier of fact to find racially discriminatory intent" in the city of Dallas's refusal to close down an illegal hazardous waste disposal site in a neighborhood of predominantly black residents (*Cox v. City of Dallas*).

Parents and the Detroit Public Schools reach an agreement on a pending case concerning the construction of a public school on an industrial site. The school system agrees to develop a system to ensure the safety of the site and a set of mechanisms permitting parents to monitor that safety (*Lucero v. Detroit Public Schools*).

2005   The EPA issues the document "Draft Final Title VI Public Involvement Guidance for EPA Assistance Recipients Administering Environmental Permitting Programs (Recipient Guidance)." The document is designed to provide guidance for personnel who administer environmental permitting programs and enhance public participation in those programs.

White residents of Dallas sue the city, claiming environmental racism in the construction of public housing in their neighborhood. They claim that the city's choice of their neighborhood for public housing was racially motivated and violated their equal protection rights under the Fourteenth Amendment. The local court disagrees with this reasoning and rejects the residents' suit, a decision that is later affirmed by the Fifth Circuit Court of Appeals (*Walker v. City of Mesquite*).

The EPA issues the "Draft Framework for Integrating Environmental Justice" and the "Draft Environmental Justice Strategic Plan Outline," which together form the foundation for the agency's Environmental Justice Strategic Plan for 2006–2011. The EPA plan is met with severe and widespread complaints, and more than 45 organizations interested in environmentalism and environmental justice sign a letter objecting to the plans.

The GAO publishes *Environmental Justice: EPA Should Devote More Attention to Environmental Justice When*

2005
(cont.)

Developing Clean Air Rules, a report on the EPA's implementation of Executive Order 12898 with regard
to air quality. The GAO concludes that the EPA has
not been sufficiently vigilant in following the provisions of the order.

The California legislature passes a bill requiring local
governments to include consideration of environmental justice issues in the development of their general plans.

New Mexico governor Bill Richardson signs an executive order requiring all state departments, boards, and
commissions to provide "meaningful opportunities"
for local communities to be involved in decisions affecting minority and low-income communities.

2006

Warren County begins plans for the construction of
Justice Park at the site of the former PCB waste disposal dump, where the environmental justice movement began in 1982.

The NRC Licensing Board rules that a claim of environmental inequity filed by neighbors of the Pal-
isades Nuclear Plant near South Haven, Michigan, is
not valid and that the plant is eligible for relicensing.

The EPA issues its "Title VI Public Involvement
Guidance for EPA Assistance Recipients Administering Environmental Permitting Programs." The purpose of the document is to assist licensing agencies in
paying proper attention to issues of environmental
justice so that they will not eventually be exposed to
Title VI claims by affected groups.

The California Global Warming Solutions Act of 2006
requires regulators to consider how emissions control
systems may affect low-income and minority communities disproportionately.

The nonpartisan think tank Resources for the Future
(RF) issues a report suggesting that grants issued by

the EPA's Office of Environmental Justice (OEJ) tend to go to communities with large minority and low-income populations, but that grant recipients tend to be exposed to relatively low concentrations of hazardous materials. RF recommends that the OEJ reevaluate the criteria under which it issues its grants to remedy this problem.

The NRC rules that five environmental justice groups objecting to the construction of the new Vogtle Electric Generating Plant near Waynesboro, Georgia, have not provided evidence that minority and low-income communities in the area will be disproportionately affected by the plant's operation.

A coalition of more than 70 organizations concerned with issues of environmental justice initiate the National Environmental Justice Tour for All, with simultaneous trips in the Northeast, the South, and the West to highlight the effects of environmental racism on poor and minority communities around the United States.

2007   An environmental justice conference is held at Howard University in Washington, D.C., "to explore new ideas and new approaches to environmental justice." A similar conference is held in 2008.

Michigan governor Jennifer Granholm signs an executive directive requiring the Michigan Department of Environmental Quality to develop and implement a state environmental justice plan that, among other things, ensures that all state agencies and departments integrate the principles of environmental justice into their decision-making procedures and activities.

A coalition of environmental justice groups in Los Angeles defeats efforts to approve the drilling of 24 new oil wells in the Baldwin Hills neighborhood of the city. Members of the same coalition had earlier prevented the construction of a power plant (2001) and a garbage dump (2003) in the area. The city finally

2007
(cont.)
agrees to build a two-square-mile park in Baldwin Hills, which has one of the most heavily concentrated populations of blacks in the city.

2008
The Oregon Task Force on Environmental Justice, created by the legislature in 2007, begins operation. Its responsibilities include ensuring that all 14 of the state's natural resources departments and commissions include procedures for input by minority, low-income, and other affected groups on issues that may involve disproportionate environmental effects on projects for which the agencies are responsible.

Sen. Hillary Rodham Clinton (D-NY) introduces the latest version of the federal environmental justice bill, S. 2549. The bill has been introduced, but never acted upon, in every legislative session for more than a decade.

5

Biographical Sketches

The biographical history of the environmental justice movement is quite different in some ways from that of other major movements, such as the traditional environmental and civil rights movements. The environmental justice movement is largely decentralized, manifested in the activities of dozens or hundreds of local groups, rather than centralized, operating under the direction of large national organizations. As a result, many of the real heroes and heroines of the environmental justice movement are organizers or chairpersons of town, county, village, or parish organizations; they are people whose names are probably not well known outside of the environmental justice movement itself.

To be fair, this chapter should probably contain biographical sketches of many of those who have led small, local groups as well as those whose names are more widely known. Since space does not permit the inclusion of all of those whose names belong here, the chapter provides a sample of women and men who have made significant contributions to the environmental justice movement, at whatever level they may have worked.

Dana Ann Alston (1951–1999)

Among her many accomplishments, Dana Ann Alston is perhaps best remembered for an inspiring speech at the 1991 First National People of Color Environmental Leadership Summit. In that speech, she challenged mainstream environmental groups to expand their view of environmental issues to include

137

the problem of environmental inequities. Modern-day leaders of the environmental justice movement continue to refer to and quote from that speech. Only a year later, Alston led a delegation of leaders from the environmental justice movement at the 1992 Earth Summit and Global Forum meeting in Rio de Janeiro, Brazil. There she spoke fervently about the international character of environmental racism and the need for environmentalists from all parts of the world to join together to deal with the problem. In 2002, the Bannerman Fellowship Program was renamed the Alston/Bannerman Fellowship Program in her honor. The program was founded in 1988 to provide support for individuals involved in organizing low-income and minority people for social causes. Alston had, herself, been a Bannerman Fellow in 1992.

Alston was born in Harlem on December 18, 1951, to Garlen and Betty Alston. She attended Wheelock College in Boston, where she served as president of the Black Student Organization and worked to increase the number of black students and faculty at the college. After receiving her BS from Wheelock in 1973, she continued her studies at the Columbia University School of Public Health, where she received her master's degree in occupational and environmental health in 1979.

One of Alston's first jobs was with Rural America, an organization formed to deal with farmworker exposure to pesticides. At the time, there was wide concern about the possible health risks posed by pesticide residues remaining on foods but relatively little attention to the hazards posed to workers in the field who were exposed to those pesticides. Partly as a result of this experience, Alston began to formulate a new understanding of environmentalism, one that does not focus as much on recreational and wildlife concerns as on the harmful effects of pollution on minority and low-income communities. Alston later held a number of positions with groups interested in environmental justice, including the National Black United Fund, where she served as president from 1985 to 1987; the Southern Rural Women's Network; the TransAfrica Forum; the Panos Institute; and the National Committee for Responsive Philanthropy. In 1991, she was recruited to serve as program officer for the environment at the Public Welfare Foundation, a philanthropic organization formed to support and finance a wide range of social groups and activities. Alston held that post until her death on

August 7, 1999, in San Francisco from kidney disease and complications from a stroke.

Bunyan Bryant (1935–)

An important conference on environmental justice was the 1990 Conference on Race and the Incidence of Environmental Hazards. It was sponsored by and held at the University of Michigan in Ann Arbor. A co-organizer of the conference was Dr. Bunyan Bryant. One consequence of the Michigan Conference was a series of high-level policy discussions with the U.S. Environmental Protection Agency (EPA) administrators William K. Reilly and Carol Browner—discussions to which Bryant was an important contributor.

Bryant has also been involved in a number of other seminal meetings and organizations from which has grown the modern environmental justice movement. For three years, he was cofacilitator of the Martin Luther King Planning Committee at the University of Michigan, where workshops on environmental justice were an integral part of celebrating Dr. King's legacy. Bryant served on the Advisory Committee of the First National People of Color Environmental Leadership Summit in 1991, was cofacilitator of the Symposium for Health Research and Needs to Ensure Environmental Justice in 1994, and was a member of the EPA's National Environmental Justice Advisory Council in 1994–1995. He has also served on dozens of boards, commissions, caucuses, committees, and task forces dealing with issues related to environmental justice.

Bryant was born in Little Rock, Arkansas, on March 6, 1935. He received his BS degree in social science from Eastern Michigan University in 1958, his MSW in social work from the University of Michigan in 1965, and his PhD in education, also from the University of Michigan, in 1970. Bryant did his postdoctoral research in town and country planning at the University of Manchester in England in 1973. He then returned to the University of Michigan as assistant professor in the School of Natural Resources and Environment. He has remained at Michigan ever since and currently holds a joint appointment as professor in the School of Natural Resources and Environment and at the Center of African-American and African Studies.

Bryant is author, coauthor, and editor of more than 60 books and articles, including *Environmental Advocacy: Concepts, Issues, and Dilemma; Race and the Incidence of Environmental Hazards: A Time for Discourse* (with Paul Mohai); *Environmental Justice: Issues, Policies, and Solutions;* and *Social Environmental Change.* He has received a number of honors and awards, including the first Damu Smith Power-of-One Environmental Justice Award of the Deep South Center for Environmental Justice (2006), the Ann Bonar Award for Gray Activism (2006), and the University of Michigan Regents' Award for Distinguished Public Service (2005).

Robert Bullard (1946–)

One of the most prolific and articulate writers in the field of environmental justice is Robert Doyle Bullard, currently Ware Distinguished Professor of sociology and director of the Environmental Justice Resource Center at Clark Atlanta University in Atlanta. Bullard's book *Dumping in Dixie* (1990, 2000, 2004) is regarded by some authorities as one of the most powerful statements on the issue of environmental justice yet to be written. Among his most recent books are *The Quest for Environmental Justice: Human Rights and the Politics of Pollution* (2005); *Growing Smarter: Achieving Livable Communities, Environmental Justice, and Regional Equity* (2007); and *The Black Metropolis in the Twenty-First Century: Race, Power, and the Politics of Place* (2007). For his many contributions to the field, Bullard is often called the father of the environmental justice movement.

Bullard was born in Elba, Alabama, on December 21, 1946. He received his bachelor's degree in government from Alabama A&M University in 1968, his MA in sociology from Atlanta University in 1972, and his PhD in the same field from Iowa State University in 1976. He also worked as an urban planner in Des Moines, Iowa (1971–1974), administrative assistant in the Office of Minority Affairs at Iowa State (1974–1975), research coordinator in Polk County, Iowa, and director of research at the Urban Research Center of Texas Southern University (1976–1978).

In 1976, Bullard was appointed assistant professor at Texas Southern University, and, in 1980, he was promoted to associate professor. He went on to hold a series of academic appointments at Rice University (1980), the University of Tennessee (1987–1988), the University of California at Berkeley (1988–1989), the University of California at Riverside (1989–1993), and the University of California at Los Angeles (1993–1994). In 1994, Bullard was appointed to his current position at Clark Atlanta University.

Bullard has twice won awards from the Gustavus Myers Center for the Study of Human Rights in North America for his books, the first for *In Search of the New South: The Black Urban Experience in the 1970s and 1980s* in 1980, and the second for *Residential Apartheid: The American Legacy* (edited with J. Eugene Grigsby III and Charles Lee) in 1996. In August 2007, Bullard was presented with the William Foote Whyte Distinguished Career Award by the American Sociological Association's Section on Sociological Practice for his contributions to the founding and development of the environmental justice movement.

Dollie Burwell (1948–)

Dollie Burwell was a leader of the protests in Warren County, North Carolina, during the late 1970s and early 1980s against the dumping of 32,000 cubic yards of soil contaminated with polychlorinated biphenyls (PCBs). When that effort was unsuccessful, she later spearheaded efforts to clean up the dump area and restore it to its natural condition, a campaign that eventually was successful. Nearly 15 years after the initial dumping in Warren County, the state of North Carolina allocated $1 million to study the toxicity of soils at the Warren County dump. As a result of that study and continued pressure by residents of the county, the state then allocated an additional $15 million for cleanup of the dump. That task was finally completed, and the former dump has become clean enough that it is now used as a recreational site.

Burwell was born in Warren County in 1948. She became interested in politics early in life and worked on a voter registration project at the age of 12. Although she came from a very poor family, she was determined to have a college education and took courses at Durham College and Shaw University. In 1971, she married William J. Burwell and moved with him to an army post at West Point. There she continued her political activities by organizing couples of color who had been denied government housing because of their race. The Burwells eventually returned to Warren County, where Dollie became involved in the dumping protest, while also raising her two young daughters. The older of those daughters, Kimberly, has also become an activist in the environmental justice movement.

Besides the four-decade battle in Warren County, Burwell has worked on other issues of environmental justice and human and civil rights problems. In addition to her private efforts, she has been involved as an aide to Rep. Eva Clayton (D-NC; 1992–2003) and, later, to Rep. G. K. Butterfield Jr. (D-NC; 2004–present). In these posts, she has served as a link between the Congress member and his or her constituents, has represented the Congress member at various events within the district, has been a caseworker in three towns in the district, and has served as the Congress member's field representative in the towns of Rocky Mount, Granville, Halifax, Person, and Vance.

César Chávez (1927–1993)

By some standards, the senior member of the environmental justice movement could be considered to be César Estrada Chávez. Decades before the Warren County protests, the Michigan Conference, or the First National People of Color Environmental Leadership Summit, Chávez was working and organizing to deal with environmental issues (as well as other kinds of problems) faced by farmworkers.

Chávez was born on a small farm near Yuma, Arizona, on March 31, 1927. His family lost its farm during the Great Depression, and they joined the exodus to California, looking for a better life. That "better life" turned out to be that of migrant workers, moving from field to field when and if work was available. Chávez attended more than 30 different schools before dropping out to continue working full time in the fields. His work there was interrupted briefly when he served in the U.S. Navy during the last two years of World War II.

Chávez's political career began in 1952, when he was invited to become first a volunteer and then a paid staff member with the Community Service Organization (CSO), an organization created by activist Saul Alinsky (1909–1972) to help the poor and politically powerless develop their own political organizations. Chávez worked his way up the ranks in the CSO until becoming general director in 1958. By that time, however, the organization was beginning to change in character, losing contact with its grassroots origins, according to Chávez. He thus resigned from the CSO in 1962, withdrew his savings from the bank, and created the National Farm Workers Association (NFWA).

The NFWA grew slowly at first as Chávez organized farmworkers one field at a time, one community after another. It recorded a few small victories in the San Joaquin and Imperial valleys of California, winning pay increases, for example, for workers in the Delano area in 1965. The major breakthrough for the NFWA came in 1968, however, when migrant grape pickers in the Delano area went on strike for higher wages. Chávez seized the opportunity to make the strike a national campaign for the workers and for the NFWA itself. He sought the aid and cooperation of major liberal establishments throughout the United States, including civil rights groups, religious organizations, major newspapers and magazines, and political leaders such as Robert F. Kennedy. The grape strike continued in one form or another for three decades, winning a number of concessions for workers along the way and establishing the NFWA (later renamed the United Farm Workers of America, or UFWA) as the most powerful representative for farmworkers, a position it held for many years.

Among Chávez's specific contributions within the UFWA were the establishment of a burial insurance program (1963); a union newspaper, *El Malcriado;* the Farm Workers Credit Union (1963); a theater group, El Teatro Campesino (1965); an artists group, El Taller Gráfico (1965); the National Farm Workers Service Center (1966); the Robert F. Kennedy Farm Worker Medical Fund (1967); the National Farm Workers Health Group (1969); Agbayani Village, a farmworker retirement community (1969); and the Juan de la Cruz Farm Workers Pension Fund (1973).

Chávez died in San Luis, Arizona, on April 23, 1993. His birthday has become a holiday in eight states—Arizona, California, Colorado, Michigan, New Mexico, Texas, Utah, and Wisconsin—and a number of schools, libraries, streets, parks, and other facilities have been named in his honor. On August 8, 1994, President Bill Clinton awarded the Presidential Medal of Freedom posthumously to Chávez.

Torri Estrada (1969–)

Torri Estrada is currently a program officer at the Marin Community Foundation, where he is responsible for determining the organization's grant awards in the area of the environment. He has a long and impressive history of work in the field of environmental justice. He earned his BA in sociocultural and ecological anthropology (1991) and his BS in environmental science, policy, and management (1992), both at the University of California at Berkeley; his AA in labor studies from the City College of San Francisco in 1994; and his MS in natural resources policy and environmental sociology from the University of Michigan in 2002.

Estrada's work résumé is extensive. It includes more than a dozen positions over the past two decades, including that of research investigator at the Pacific Energy and Resources Center in Sausalito, California; environmental educator with the East Bay Municipal Utility District in Oakland, California; research analyst for the Sierra Club Legal Defense Fund in San Francisco; mediation consultant at the Federal Mediation and Conciliation Service in Oakland; project investigator at the Native American Renewable Energy Evaluation Project in Berkeley, California; project director at the Urban Habitat Program in San Francisco; consultant at the Environmental Justice Coalition on Water in Oakland; senior policy fellow at the Latino Issues Form in San Francisco; and program officer at the Unitarian Universalist Veatch Program at Shelter Rock in Manhasset, New York.

Estrada was born in Montclair, California, on November 13, 1969. He has been honored with a School of Natural Resources and Environment Merit Award (1995–1997), a Rackham Merit Fellowship from the University of Michigan (1995–1998), a Water Education Foundation Fellowship (1998–1999), an Environmental Leadership Fellowship (1999–2002), and a New Voices Fellowship (2001–2003).

Margie Eugene-Richard (dates unknown)

Margie Eugene-Richard is a longtime activist in the environmental justice movement, working on health issues arising out of exposure to hazardous materials in Louisiana's notorious "Cancer Alley."

In the 1970s, Eugene-Richard was a middle school teacher living in Norco, Louisiana, in Cancer Alley. The region's name came from the extensive land, air, and water pollution produced by numerous petroleum and chemical plants in the area. Eugene-Richard, in fact, had grown up in a house located less than 25 feet from a 15-acre Shell Oil chemical plant. The predominance of industry in the area is reflected in the name of the town itself, an acronym for the New Orleans Refinery Company. The Old Diamond community in which Eugene-Richard grew up is trapped between Shell refineries on two sides. Residents have long been accustomed to explosions, toxic spills, and all forms of pollution from the two facilities.

Eugene-Richard says that a defining event in her life occurred in 1973, when an exploding pipeline blew one house off its foundations and killed an elderly woman and a young boy. From that point on, she realized that community action was necessary to combat the dangers posed by Shell and other industrial operations in the area. She planned workshops for local citizens and developed methods for monitoring pollution in the community. One especially effective program was the Bucket Brigade, a system by which local citizens could capture air pollutants in "buckets," allowing them to identify and quantify the environmental risks to which they were being exposed. In 1989, Eugene-Richard founded Concerned Citizens of Norco (CCN) to strengthen and formalize the battles against Shell.

Those battles went on for more than a decade before the company agreed to a number of actions to reduce its impact on the surrounding environment. It agreed to pay relocation costs for citizens of Norco closest to its outer boundaries, to reduce its emissions by 30 percent, to contribute $5 million to a community development fund, and to develop an emergency evacuation plan and system in case of a major disaster. Since the Shell agreement was signed in 2000, the company has bought up more than 200 of the 225 residential lots in Norco.

In addition to her work with CCN, Eugene-Richard has testified at hearings before the U.S. Congress and the UN Human Rights Commission and has spoken before the World Conference against Racism in 2001 and the World Summit on Sustainable Development in 2002. In 2004, she became the first African American to win the prestigious Goldman Environmental Prize honoring grassroots environmental heroes. The prize carries with it a cash award of $150,000 and a bronze sculpture called the Ouroboros.

Jay Feldman (1953–)

Jay Feldman has been a prominent spokesman for issues involving farmworker health issues for more than three decades.

Feldman was one of the founders of the National Coalition against the Misuse of Pesticides (NCAMP) in 1981 and has served as its executive director ever since. In 1998, NCAMP changed its name to Beyond Pesticides to better reflect the organization's mission and goals. In his role as executive director, Feldman provides policy direction and overall coordination of Beyond Pesticide's activities in controlling the use of pesticides and promoting nonsynthetic chemical alternatives. He is also editor of the organization's newsletter, "Pesticides and You."

Jay Feldman was born in Brooklyn, New York, on May 9, 1953. He earned his BA in political science from Grinnell College, in Grinnell, Iowa, in 1975 and his MA in urban and regional planning from Virginia Polytechnic Institute (now Virginia Polytechnic Institute and State University) in 1977. In 1974, he was a researcher for Associated Colleges of the Midwest in San Jose, Costa Rica. From 1977 to 1979, Feldman worked as a rural health specialist at Rural America, a national advocacy organization for people living in small towns and rural areas, and from 1979 to 1981 he was director of the organization's health programs.

Among Feldman's publications are *Rural Health Directory: A Resource Guide to Nongovernmental Organizations Involved in Rural Health* (1978), *Pesticide Use and Misuse: Farmworkers and Small Farmers Speak on the Problem* (1980), and *Voices for Pesticide Reform: The Case for Safe Practices and Sound Policy* (with Sandra Schubert and Terry Shistar, editors, 1996). He has also written for a number of newspapers and journals, including the *New York Times*, *Washington Post, Los Angeles Times,* and *Environmental Law Reporter.* Feldman is also a popular guest on a number of radio and television programs, including *Nightline, The NewsHour with Jim Lehrer, Good Morning America,* and *The Today Show.*

Deeohn Ferris (1953–)

For nearly three decades, Deeohn Ferris has worked on issues of environmental law. During the last quarter of that period, she has focused on the topic of environmental justice.

Ferris was born on November 3, 1953, in Norwalk, Ohio. She received her BA from Ashland University in Ashland, Ohio, and her JD from Georgetown University Law Center in 1978. Her first job was with the EPA, where she worked as attorney-advisor in the Office of Legislation (1979–1983), as assistant enforcement counsel in the Office of Enforcement and Compliance Monitoring (1983–1984), and as director of the Special Litigation Division (1984–1986). Her key accomplishments at EPA included handling civil litigation aimed at fostering compliance with chemical assessment and control programs, implementing the first full-scale environmental and management audits, and demonstrating effective new federal compliance tools.

From 1986 to 1989, Ferris worked as environmental counsel on liability and toxic tort issues at the American Insurance Association. She then moved to the National Wildlife Federation (NWF) as director of environmental quality from 1990 to 1992. At NWF, she published a quarterly newsletter, "The Gene Exchanged"; coauthored a national report, *Waters at Risk: Keeping Clean Waters Clean;* and was awarded the Charlie Shaw Award for exceptional contributions to the shared mission of NWF and its affiliates.

In 1992, Ferris accepted a position as program director of the Environmental Justice Project at the Lawyer's Committee for Civil Rights under Law. While there, she organized and launched a national environmental justice project that provides legal and technical assistance to clients and advocates for legislation and regulation. During this period, she was also appointed to the EPA Federal Environmental Justice Advisory Committee. In 1993, Ferris conceptualized the executive order on environmental justice issued by President Bill Clinton in 1994.

In 1994, Ferris was appointed director of the newly created Alliance for the Washington Office on Environmental Justice, an international collaborative of community-based regional networks and organizations. She is currently president of both the Sustainable Community Development Group, Inc., a not-for-profit corporation dedicated to metropolitan sustainability, environmental health, smart growth, and regional equity, and Global Environmental Resources, Inc., working to "help clients reach a goal of sustainable development by taking into account the causes and effects of environmental, health and economic factors." Ferris is a popular speaker on law, public health, and policy and has presented addresses and worked with communities in Brazil, China, Fiji, India, Indonesia, Mexico, Nepal, South Africa, Turkey, and Mexico.

LaDonna Harris (1931–)

For more than 40 years, LaDonna Vita Tabbytite Harris has been an advocate for the rights of Native Americans. One of her earliest accomplishments was the founding in 1965 of Oklahomans for Indian Opportunity (OIO), the first intertribal organization in the state. In 1970, she established a national organization similar to OIO, Americans for Indian Opportunity (AIO), an organization for which she has served as president for nearly four decades.

Harris was born in Temple, Oklahoma, on February 15, 1931, to Lily Tabbytite, a member of the Comanche tribe, and Donald Crawford, of Irish heritage. Her parents separated when LaDonna was still a young child, and she was raised by her maternal grandparents on a small farm outside Walters, Oklahoma. She spoke only her native Comanche language until she entered elementary school. Since she inherited some of her father's white skin color, she experienced less racism while growing up than did most Native Americans at the time. She had no formal education beyond high school, from which she graduated in 1949, but she has received several honorary degrees from institutions such as the University of Oklahoma, Northern Michigan University, Antioch College, Dartmouth College, and Marymount College of New York. While still a senior in high school, she married her high school sweetheart, Fred Harris, who was later to serve in the Oklahoma State Senate at the age of 25. Harris was also elected to the U.S. Senate from Oklahoma in 1964. He served in the Senate until 1972, when he decided to run for president of the United States, a campaign that was eventually unsuccessful.

In addition to her work with OIO and AIO, Harris has been involved in the establishment of the Council for Energy Resource Tribes, the National Women's Political Caucus, the National Indian Housing Council, the National Indian Business Association, and the National Tribal Council on the Environment. In recent years, much of Harris's work has focused on the AIO's Ambassador's Program, designed to foster leadership growth within Native American communities. That program has now been extended to include indigenous communities in other parts of the world, such as New Zealand and Bolivia.

Among the many awards Harris has received are the Outstanding American Citizen Award, the United Nations Peace Medal, the *Ladies' Home Journal* Woman of the Year Award, the

Human Rights Award of Delta Sigma Theta Sorority, and the Award for Public Service of Theta Sigma Phi, National Journalism Fraternity for Women. Harris's publications include *To Govern and to Be Governed: American Indian Tribal Governments at the Crossroads* (1983), *Survey of American Indian Environmental Protection Needs on Reservation Lands* (1986), *Partnerships for the Protection of Tribal Environments* (1991), and *LaDonna Harris: A Comanche Life* (2000). In 1980, Harris was nominated for vice president of the United States by the short-lived Citizens Party.

Hazel Johnson (dates unknown)

In 1969, Hazel Johnson's husband died of lung cancer at the age of 41. Instead of just acknowledging her sorrow and going on with her life, Johnson began to question the cause of death of such a young man. The more she asked, the more she learned about the unhealthy housing development—Altgeld Gardens—in which she lived. She soon found that Altgeld Gardens was at the center of a "toxic doughnut," with the highest incidence of cancer of any neighborhood in Chicago. The toxic doughnut consisted of a ring more than 140 miles in circumference of incinerators, water and sewage treatment plants, steel mills, paint factories, scrap yards, more than 50 landfills, and 3 abandoned wastewater lagoons. The effluents from these facilities appeared to be responsible for the astonishing number of health problems—asthma and other allergies, excessive nosebleeds and headaches, dizziness and fainting spells, and high rates of cancer—recorded for Altgeld Gardens.

By 1979, Johnson had begun to organize her community to learn more about the health problems they faced and to do something about those problems. The organization she formed, People for Community Recovery (PCR), was eventually incorporated in 1982. Conducting door-to-door surveys of health issues among local residents, PCR volunteers found that more than 90 percent of the 10,000 residents of Altgeld Gardens reported one health problem or more. When Johnson presented these results to city officials, they rejected pending plans for the construction of a new chemical waste incinerator in the area and decided not to expand existing landfills, as they had planned to do.

Johnson and the PCR have accomplished a number of other successes. They organized a campaign to have asbestos removed from resident apartments and local schools, developed a program on lead poisoning in the community, created a special study group to deal with a sodium oxide spill in the community, monitored the cleanup of PCB contamination in the area, and sponsored a number of regional conferences on environmental justice issues. Today, Johnson has ceded many leadership responsibilities at PCR to her daughter, Cheryl.

For her many accomplishments, Hazel Johnson was named the "mother of the environmental justice movement" at the First National People of Color Environmental Leadership Summit in 1991. In December 2006, Illinois lieutenant governor Pat Quinn awarded Johnson a state Environmental Hero Award in recognition of her commitment to environmental health and protection. Johnson has also accepted the President's Environment and Conservation Challenge Award on behalf of PCR from President George H. W. Bush.

Pamela Tau Lee (1948–)

Pamela Tau Lee has been coordinator of public programs in the Labor Occupational Health Program (LOHP) in the School of Public Health at the University of California at Berkeley since 1990. In that position, she is responsible for training, technical assistance, materials development, and planning of conferences on occupational health issues. She also coordinates outreach efforts to labor, community, and environmental justice groups.

Lee was born in San Francisco in 1948. She attended California State University at Hayward (CSU-H; now California State University, East Bay), from which she received her bachelor's degree in sociology in 1969. She then attended a Teacher Corps training program at the Secondary Education Teacher Post Graduate Program at CSU-H. (The Teacher Corps was created in 1965 to provide teachers for underserved low-income areas throughout the nation.) As part of this program, she did her student teaching in nearby Oakland, where she was first introduced to problems faced by inner-city minority adults and children. She later said that the experience was very influential in her life because "[t]here was so much activism then." As a result, she herself became involved in efforts to identify and deal with the issues faced by inner-city minorities.

After completing her Teacher Corps training, Lee returned to San Francisco, where she worked as a volunteer for the Chinese Progressive Organization. There she saw firsthand the health issues endemic among members of the Asian American community, who were often required to work at backbreaking jobs in environmentally hazardous conditions. In connection with her own job as a hotel maid, she eventually became involved as a job steward with the Hotel Employees/Restaurant Employees Union Local 2 AFL-CIO in San Francisco. Over time, she was promoted to higher positions until she was named staff director of the union in 1985. She held that post until moving to LOHP in 1990.

Since 1990, Lee has been involved in a number of local, regional, and national conferences on environmental justice. At the 1991 People of Color Environmental Leadership Summit, she and about 30 other Asian and Pacific Island Americans realized that a formal organization was needed to deal with specific issues related to their communities. Two years later, that concern was expressed in the formation of the Asian Pacific Environmental Network (APEN), for which she has served in a number of capacities. In 2005, she resigned as chair of the board of directors of APEN because of term limits. She is currently a member of the National Environmental Justice Advisory Council of the EPA and of the National Liaison Committee for the National Occupational Research Agenda (NORA) and the National Institute for Occupational Safety and Health (NIOSH). Among her awards are the Golden Apple, Distinguished Community Member for Excellence Award of the University of California School of Public Health; the Lorin E. Kerr Award of the American Public Health Association; the Changing Face of Women's Health Recognition Award of the National Health Sciences Consortium; and the Bread and Roses Recognition Award of the Second People of Color Environmental Leadership Summit.

Graciela Martinez (1945–)

Graciela Martinez learned about environmental issues early in her life. When she was still a young girl, her mother, a farmworker, often took her into the gladiolus fields where she worked. When she did so, she made her wear long pants, even though it was very hot. After finishing work every day, she also had to wash those pants and all the other clothes she wore in the fields very thoroughly. Her mother explained that these actions were necessary to remove all traces of the pesticides used on the flowers, pesticides that could cause serious health problems for workers in the fields.

Martinez was born on January 29, 1945, in Harlingen, Texas, to Paula and José Martinez. José was a stonemason, and Paula, a housewife, in addition to her work in the fields. The family moved to Farmersville, California, in 1955, where Graciela graduated from high school. Her first job was as a typist for the American Friends Service Committee (AFSC). That job did not last long, however, as Martinez was offered an opportunity to work as secretary to labor organizer and activist César Chávez, a position that allowed her to become involved in her true love, the environmental justice movement. Martinez remained with Chávez's United Farm Workers for a number of years as both a paid and volunteer worker. She was also involved in a number of environmental, civil rights, and community organizations and activities.

For the past decade, Martinez has been employed as program coordinator for the AFSC's Proyecto Campesino program in its Visalia, California, office. In that position, she is responsible for organizing tenant associations, assisting with immigration and naturalization issues, supporting local community organizing efforts, and advocating for labor and human rights. She also hosts an AFSC radio program, *The Voice of the Community*, that deals with issues of interest to local listeners, such as water use and pollution, health, immigration and naturalization, and unfair labor practices. In 2006, Martinez was awarded the 15th Annual Dorothy Richardson Leadership Award of the Institute for Developing Emerging Area Leaders (IDEAL), a program from which she was a graduate in 1999.

Norris McDonald (1953–)

Norris McDonald is president of the African American Environmentalist Association (AAEA), a group he founded in 1985. He saw a need for a group consisting primarily of people of color interested in environmental issues, because most existing environmental groups were largely made up of whites. He felt that people of color did not feel they had a vested interest for dealing with critical environmental problems that had a direct impact on their lives. Today, AAEA engages in a wide variety of activities to educate people of color, such as conducting tours of toxic sites, power plants, water purification and sewage treatment plants, and conservation facilities. The organization also lobbies local and federal decision-making agencies and cooperates with a number of other organizations interested in environmental justice.

McDonald was born in Thomasville, North Carolina, on February 8, 1953, to Sandy Norris McDonald Sr. and Katie Louvenia Best, a high school principal and an assistant in the public school system, respectively. He attended Wake Forest University, from which he received his bachelor's degree in education in 1977. Shortly thereafter, an event occurred that was to have a crucial impact on his life: the accident at the Three Mile Island nuclear power plant near Middletown, Pennsylvania, on March 28, 1979. McDonald was struck by the magnitude of the environmental problems facing the nation and decided to make a career in the environmental movement. A few months after the Three Mile Island event, he joined the Environmental Policy Center (EPC; now Friends of the Earth), where he became director of the Energy Conservation and Transportation Project. A series of personal events only strengthened McDonald's decision to focus on environmental issues. In 1991, he developed chronic, acute asthma, which led to two life-threatening episodes in July of that year and again in 1996. He is convinced that his health problems are directly attributable to air pollution. In 1985, McDonald left EPC to found AAEA.

Paul Mohai (1949–)

One of the most active researchers in the field of environmental inequities has been Paul Mohai. For more than two decades, Mohai has analyzed the social, political, and organizational processes that influence and shape environmental and natural resource policy. He has also explored public attitudes toward environmental and natural resource issues, analyzing the factors that affect political activism, assessing the impact of environmental pollution and resource scarcities on low-income and minority groups, and evaluating the contribution of wildland recreation to the quality of life.

Mohai was born on June 20, 1949, in Linz, Austria. For his undergraduate education, he attended Michigan State University (1967–1968), the University of Birmingham, England (1969–1970), and the University of California at Berkeley (1968–1971), from which he received his BA in mathematics. He then earned his MS degree in forestry and statistics from Syracuse University in 1976 and his PhD in natural resource policy and the sociology of natural resources from Pennsylvania State University in 1983.

Mohai's first professional appointment was as mathematics teacher at Fox Lane Middle School in Bedford, New York. He then held assignments as teaching assistant and instructor at the State University of New York College of Environmental Science and Forestry at Syracuse (1973–1976), as project intern and program assistant at the Natural Resources and Forestry Group of the U.S. Department of Agriculture's Cooperative State Research Service (1977–1979), as instructor in the School of Forest Resources at Penn State (1979–1983), and as assistant professor in the College of Natural Resources, Utah State University (1983–1987).

In 1987, Mohai accepted an appointment in the School of Natural Resources and Environment at the University of Michigan. He now holds a professorship in the School of Natural Resources and Environment and is a faculty associate in the Social Environment and Health Program of the Survey Research Center of the Institute for Social Research at Michigan. During the summer of 1995, Mohai was visiting professor in the Department of Rural Sociology at the University of Wisconsin, Madison.

Mohai has published more than 40 papers in peer-reviewed journals and made more than 40 presentations at various conferences and meetings. He has testified twice before congressional committees on environmental justice issues. His most recent publication is *Toxic Wastes and Race at Twenty, 1987–2007: Grassroots Struggles to Dismantle Environmental Racism in the United States* (with Robert D. Bullard, Robin Saha, and Beverly Wright; 2007). Mohai currently serves as a representative of academia on the National Environmental Justice Advisory Council of the EPA.

Richard Moore (dates unknown)

Richard Moore has been involved in a variety of social issues for more than 40 years. He has worked with community-based groups around issues such as welfare rights, police brutality, street gang activities, drug abuse, low-cost health care, child nutrition, and the fight against racism. He has been active in the environmental justice movement from its earliest days, serving on the planning committee for the First National People of Color Environmental Leadership Summit in October 1991. In 1969, while a member of the Black Beret organization centered in the San Jose district of Albuquerque, Moore founded the Bobby Garcia Memorial Clinic, named in honor of a local community organizer and activist. He was also a founding member of the Southwest Organizing Project (SWOP), a multiracial organization working to empower the disenfranchised in the Southwest, and of the Southwest Network for Environmental and Economic Justice (SNEEJ), a regional organization consisting of more than 80 community-based grassroots organizations working in communities of color in six southwestern states and Mexico. Moore currently serves as executive director of SNEEJ. In 2005, he was awarded the Ford Foundation Leadership for a Changing World Award.

Moore credits his interest in social issues to his upbringing in a poor, single-parent family in Albuquerque. The family home was adjacent to the city's sewage treatment plant, and he saw firsthand the offensive and dangerous effects of the facility on the neighborhood. He dropped out of high school at the age of 16 convinced that he wanted to do what he could to make a change in his community. He knew, he later said, that "things should not be this way, and don't need to be this way." Much of his early work was couched not in terms of "environmental justice," he notes, but as a form of social justice. Through his work, Moore has woven a whole range of issues, including both social and environmental issues, into a single campaign to improve the lives of poor and minority communities.

Na'taki Osborne (1974–)

Na'taki Osborne had an early introduction to the issue of environmental justice. She was born in Baton Rouge, Louisiana, not far from the state's notorious Cancer Alley. The region is so named because experts believe the high rates of cancer and other diseases observed in the region are caused by toxic chemicals released by petroleum refineries, chemical plants, and other industries in the area. The potential threat posed by Cancer Alley was brought home to Osborne when her mother, then an English professor, was diagnosed with breast cancer. Osborne could not but wonder if there was a connection between her mother's health problems and the effluents from chemical plants in the area. She decided that the study of environmental problems and actions for dealing with such problems were to be an important part of her life.

Osborne attended Spelman College in Atlanta, from which she earned her BA degree in chemistry in 1996. Concurrently, she received a BS degree in civil and environmental engineering from Georgia Institute of Technology. She continued her studies at the Rollins School of Public Health at Emory University in Atlanta, earning her master's degree in environmental and occupational health in 2002. She is currently enrolled in a doctoral program at Emory University.

Osborne's first work experience was as an intern in the Superfund and Hazardous Waste Division of the EPA in 1994. Her job description called for her to survey people living near waste sites, identifying existing and potential health problems and developing methods for preventing or ameliorating those problems by cleaning up the waste sites. She also worked as an intern for the National Wildlife Federation (NWF) during her undergraduate years at Spelman. After graduation, she took a paid full-time job at NWF working with a minority community in Atlanta on issues of environmental education, environmental justice, community organizing, watershed and green space protection, environmental health, and related issues.

Osborne is currently the national leadership development coordinator for NWF, with responsibility for developing urban environmental initiatives and building partnerships among grassroots community organizations, academic institutions, local and state environmental agencies, and health agencies for dealing with problems such as watershed protection; land-use and habitat protection and restoration; and the effects of sprawl on plant, animal, and human communities. She is also a senior fellow at the Environmental Leadership Program in Washington, D.C., and a community health team teacher and guest environmental health professional at the Morehouse School of Medicine in Atlanta.

Swati Prakash (1975–)

Swati Prakash was formerly the director of Environmental Health and Community-Based Research at West Harlem Environmental Action (WE ACT). Her responsibilities there included the education and empowerment of community residents for addressing environmental health issues, the promotion of community-driven environmental health research, and the conduct of air pollution monitoring and reduction projects. In January 2006, Prakash left WE ACT to become program director at the Pacific Institute for Studies in Development, Environment, and Security in Oakland, California.

Prakash's early life was not one that might suggest a later involvement in the environmental justice movement. She was born on August 7, 1975, to Indian immigrants Jaya and Madhavi Prakash, an electrical engineer and administrator, respectively. She grew up in a white middle-class neighborhood in suburban New Jersey with one serious drawback: It was located near a number of petrochemical refineries, incinerators, and industrial plants for which New Jersey is famous. Later in life, she learned that the high school she had attended was built on top of a medical waste dump.

Her awareness of environmental justice issues came slowly. During high school, she became involved in both the Student Environmental Action Coalition (SEAC) and the Green Party, as well as in an antiracism club, Students Helping Overcome Tensions. She eventually realized that she was the only student engaged in both environmental and racism issues in her school. Her efforts to get students from both groups working together, however, were largely a failure.

An important turning point in Prakash's life came in 1993, when she attended a national conference of SEAC held in Berkeley, California. She was shocked to discover that, in contrast to her experience in New Jersey, virtually every person at the meeting was a person of color. For the first time in her life, the connection between race and environmentalism became clear. As one of her biographers has written, "It was the first time she felt that she did not have to trade off her identity as a South Asian person with her desire to be an environmentalist."

Prakash holds a bachelor's degree in environmental science and public policy from Harvard University and a master's degree in environmental health from Harvard's School of Public Health.

Patricia E. Salkin (dates unknown)

Patricia E. Salkin has long been involved in issues of land-use law, especially as they relate to problems of environmental racism and environmental justice. She is currently Raymond and Ella Smith Distinguished Professor of Law, associate dean, and director of the Government Law Center at Albany Law School in Albany, New York.

Salkin received her BA from the University at Albany and her JD from Albany Law School. Prior to joining the Albany Law faculty in 1992, she was an intern in the office of Rep. Benjamin A. Gilman (R-NY); legal aide for the New York State Department of State, Office of Local Government Services and Coastal Zone Management Program; law clerk at Ruthman, Feinberg & Dumas in Albany; and assistant counsel for the New York State Office of Rural Affairs. In addition to her faculty appointment at Albany Law, Salkin has taught at Sage Junior College in Albany (1989), the Rensselaer Polytechnic Institute Center for Urban and Environmental Studies (1990–1992), the University at Albany (1993–present), and the University of Pennsylvania's School of Design (2007). She is currently a member of the National Environmental Justice Advisory Council of the EPA.

Salkin has written more than two dozen books, textbooks, and treatises on land use and related topics. She has also written more than 50 law articles and columns and more than three-dozen articles for other scholarly publications. Among the many honors she has been accorded are the Outstanding Young Alumna Award, Albany Law School (1996); the Bernard E. Harvith Award for Environmental Law of the Albany Law School Environmental Alumni (2000); the Award for Excellence in Public Service of the New York State Bar Association's Committee on Attorneys in Public Service (2002); and the Distinguished Member Award of the Capital District Women's Bar Association (2007).

Peggy Shepard (1946–)

In 1988, Peggy Shepard cofounded (with Vernice Miller) West Harlem Environmental Action (WE ACT), an organization of which she is still executive director.

Shepard was born on September 16, 1946, in Lawrenceville, New Jersey. She received her BA degree in English from Howard University in 1967 and continued her studies in management at Baruch College in New York City from 1983 to 1985. Over the last two decades, she has had two careers, first as a journalist and then as a political activist. In her first career, she became the first African American reporter at the *Indianapolis News* in 1969, where she was later home furnishings editor for the paper. She then moved on to a number of other positions, including text and photo researcher for Time-Life Books (1971–1973), copy editor for the *San Juan Star* (1974–1975), associate editor at *Redbook* (1976–1977), and editorial director of *Verve* (1977–1978).

In 1978, Shepard began a long career in government-related work when she accepted a position as consumer affairs specialist at the New York State Division of Housing and Community Renewal. She later became consumer affairs director at the division (1980–1983), special assistant to the commissioner (1983–1985), director of public information for rent administration (1985–1988), and special assistant for government and community affairs (1988–1993). She was then appointed women's coordinator for the New York City Office of the Comptroller.

Shepard's life took a dramatic turn in 1988, when she became involved in efforts to prevent the construction of the North River sewage treatment plant and a Metro Transit Authority bus stop in north Manhattan. During a protest demonstration, she and six other community organizers were arrested, an action that only confirmed her commitment to work for environmental justice. Later in the year, she and Vernice Miller joined to establish WE ACT to work against environmental racism in the community. The organization was the first African American and Hispanic American community-based group for educating about and mobilizing residents around environmental health and quality-of-life issues exclusively. WE ACT was later party to a suit against New York City that resulted in an odor abatement addition to the sewage treatment plant and a $1.1 million settlement fund for the West Harlem community.

In addition to her current position as executive director of WE ACT, Shepard has held a number of other posts, including president of the Manhattan chapter of the National Women's Political Caucus from 1993 to 1997 and cochair of the Northeast Environmental Justice Network. In 2002, she was elected the first woman chair of the National Environmental Justice Advisory Council of the EPA. Shepard also serves on a number of other boards and committees dealing with environmental issues. Among the many honors and awards she has received are the New Yorker for New York Award from the Citizens Committee of New York (1990), a People Who Make a Difference Award from *National Wildlife* magazine (1993), the Earth Day Award for Excellence in Environmental Advocacy from Earth Day New York (1998), the 10th Annual Heinz Award for the Environment from the Heinz Foundation (2003), and the Dean's Distinguished Service Award from Columbia University's School of Public Health (2004).

Gail Small (dates unknown)

Issues of environmental racism affect every minority community in the United States: African American, Hispanic American, Asian American, and Native American, to name only the most prominent groups. Every community owes progress in the field of environmental justice to a number of leaders who have spoken and worked to combat the injustices their people have encountered. One of the longtime leaders of the Native American environmental justice communities is Gail Small, whose Cheyenne name is Head Chief Woman.

Small became involved in environmental justice while still a teenager in Montana. She became involved in the infamous Montana Coal Wars, initiated as a result of efforts by coal companies and the federal government to conduct extensive mining under reservation lands. She recalls that, at the time, she was the youngest person (at the age of 21) to be involved in the committee working to get coal leases canceled. As with many environmental issues, the dispute between coal companies and tribes was not easily resolved. Indeed, it was another 15 years before the companies agreed to discontinue their operations on reservation lands.

By that time, Small had finished high school, gone on to college, and earned her BA degree from the University of Montana (1978) and her JD and Certificate of Completion in Environmental and Natural Resource Law from the University of Oregon School of Law (1982). She then returned to the Northern Cheyenne reservation in southeastern Montana to discover that there was no work available for her. To resolve that problem and to continue her efforts against environmental racism, Small and a group of friends founded Native Action, one of the first nonprofit organizations established on a Native American reservation. Native Action has taken a number of steps toward improving the quality of reservation life, including the creation of the first bank, the first public high school, and the first chamber of commerce on the reservation; the development of a tribal environmental policy; a campaign against Native American voter discrimination; and the creation of a number of tribal laws dealing with community issues such as traditional burial rites, sexual assault, and domestic violence and of a uniform commercial code for small business operations.

Environmental issues remain at the top of Small's agenda. Most recently, she has been campaigning against plans to bore as many as 75,000 methane wells in areas around the reservation. The pollution produced by those wells, Small says, would turn the reservation into a wasteland.

Small has received a number of honors and awards for her work, including a 1995 *Ms. Magazine* Gloria Steinem Women of Vision Award and a 1997 Territory Resource Foundation Jeanette Rankin Award. She has served as an elected member of the Northern Cheyenne Tribal Council; has taught natural resource law and federal Indian law at the University of California at Humboldt, Chief Dull Knife Memorial College, and Little Big Horn Community College; has testified before congressional oversight hearings; and has served on the Federal Reserve Board's Consumer Advisory Council. She is also the mother of four children.

Damu Smith (1952–2006)

Environmental justice is only one of many causes to which Damu Smith devoted his life. Born in St. Louis, on December 6, 1952 (some sources say 1951), Smith was given the baptismal name of Leroy Wesley. He later changed his name to Damu Amiri Imara Smith. In Swahili, his new name means "blood, leadership, and strength"; Smith said that he chose the name because he was willing to shed his own blood for the liberation of his people. Smith's father was a fireman and an air pollution inspector, and his mother, a licensed practical nurse. The family often struggled to make ends meet, and Smith later wrote that he knew what it was like to study by candlelight, survive on food stamps and government handouts, and go to schools that had no heat. As a result, he said, he had "great sensitivity to the plight of poor people."

A seminal event in Smith's life occurred while he was still in high school. He was invited to an after-school program for disadvantaged young men that involved a field trip to Cairo, Illinois, for Black Solidarity Day. On the trip, he toured African American neighborhoods where houses had been sprayed with gunfire by white supremacists. Damu later noted that the experience of seeing those bullet holes changed his life. He developed a new appreciation for the political issues important in his life and a commitment to working on those issues.

After completing high school, Smith enrolled at St. John's University in Collegeville, Minnesota, where he served as president of the Organization of Afro-American Students. In that post, he led a protest and takeover of the school's administrative offices, demanding a black studies program. In 1973, Smith moved to Washington, D.C., to attend Antioch College's Center for the Study of Basic Human Problems and, as he said, "to be closer to the action." For Smith, that "action" involved work with a wide variety of organizations, including the United Church of Christ Commission for Racial Justice, Black Voices for Peace, the American Friends Service Committee, the Jesse Jackson presidential campaign of 1984, Greenpeace USA, the National Black Independent Political Party, the National Alliance against Racist and Political Repression, and the National Wilmington 10 Defense Committee.

In 1991, Smith became the first coordinator for environmental justice for the Southern Organizing Committee for Economic and Social Justice. In that capacity, he visited more than 40 towns and cities to view the effects of chemical dumping on minority communities. He later drew on that experience as national associate director for Greenpeace by organizing a series of Toxic Tours in the South, in which he brought celebrities to Cancer Alley in Louisiana.

In December 1999, Smith and about 200 other black community leaders and activists met in New Orleans to form the Interim National Black Environmental & Economic Justice Coordinating Committee, later renamed the National Black Environmental Justice Network (NBEJN). The organization's mission is to "enable African Americans to secure environmental and economic justice, health equity, political power, and civil rights and human rights." Smith served as executive director of NBEJN until his death from colorectal cancer on May 5, 2006, in Washington, D.C.

Beverly Wright (1947–)

Beverly Hendrix Wright has been engaged over the past three decades in a number of activities, including as a member of the Michigan Coalition, the National Advisory Committee for the First National People of Color Environmental Leadership Summit, the Planning and Protocol committees for the National Institute for Environmental Health Sciences' Health and Research Needs to Ensure Environmental Justice National Symposium, and the EPA's National Environmental Justice Advisory Council, as well as being founder and director of the Deep South Center for Environmental Justice. She has also been active in other professional activities sponsored by the National Council for Negro Women, the Association of Black Psychologists, the Association of Social and Behavioral Scientists, and the Society for the Study of Social Problems.

Wright was born in New Orleans on October 1, 1947. She earned her BA degree in sociology from Grambling College in 1969, and her MA (1971) and PhD (1977), both in sociology, from the State University of New York at Buffalo. She served as instructor in the department of sociology at Millard Fillmore College (1970–1971) and at the State University of New York at Buffalo (1970–1974). Wright was appointed to the faculty at the University of New Orleans in 1974 and was promoted to assistant professor in 1977 and to associate professor in 1987. During the period from 1989 to 1993, she served as associate professor of sociology at Wake Forest University. In 1992, she founded the Deep South Center for Environmental Justice (DSCEJ) at Xavier University of Louisiana in New Orleans. She was named director of the center, a post she continues to hold.

In 2005, DSCEJ made plans to move its operations from Xavier to Dillard University, across town in New Orleans. That move was scheduled to take place on September 1 but was disrupted when Hurricane Katrina struck the city three days earlier. In spite of the chaos and destruction caused by the hurricane, the move eventually took place, and the center is now at its new home at Dillard. One of the major projects on which it has focused its attention in the past few years has been assisting in the city's recovery from one of the nation's greatest natural disasters.

In addition to her work at DSCEJ, Wright has served on a number of community and professional committees and commissions, including the New Orleans Mayor's Environmental Advisory Committee, the Mayor's Office of Environmental Affairs Brownfields Consortium, the New Orleans Mayor's Committee on Solid Waste, the Army Corps of Engineers' Environmental Advisory Board, and the U.S. Commission of Civil Rights for the state of Louisiana. She is the author or coauthor of more than two dozen papers, reviews, and book chapters. She has received a number of awards and honors, most recently a Special Gulf Coast Award for outstanding leadership in the aftermath of Hurricane Katrina from the Robert Wood Johnson Community Health Leadership Program in 2006.

6

Data and Documents

Legislative and legal issues are an important part of the environmental justice movement. To a large extent, obtaining redress for environmental inequities and preventing such inequities from occurring in the future mean that people must know about existing environmental and civil rights laws and know how to use those laws. It also means having new laws and regulations written that will accomplish these same goals.

Early regulation and legislation dealing specifically with environmental justice date to less than two decades ago. In 1994, President Bill Clinton issued Executive Order 12898, requiring that all federal agencies develop a policy for dealing with environmental justice. In the same year, the state of New Hampshire became the first state to issue a similar requirement of all its agencies. Currently, 32 states and the District of Columbia have adopted some type of environmental justice law, regulation, or policy. Ten additional states have informal policies about environmental justice and/or individuals or offices with responsibility for enforcing environmental equity for all people in the state. An excellent source of information about state laws, regulations, and policies is *Environmental Justice for All: A Fifty State Survey of Legislation, Policies and Cases*, 3rd edition, Public Law Research Institute, University of California Hastings College of the Law, 2007.

Federal legislation on environmental justice has been very limited. In 2008, Sen. Hillary Rodham Clinton (D-NY) introduced legislation establishing the Interagency Working Group on Environmental Justice to provide guidance to all federal agencies on issues related to environmental justice. Other than Clinton's bill, there has been almost no federal legislation dealing

165

specifically with environmental justice. Nonetheless, legislative justification for many principles of environmental justice can be found in a number of more general laws dealing with civil rights and environmental issues. This chapter includes a number of laws, bills, executive orders, regulations, and policy statements dealing with environmental justice. It also includes portions of the Basel Convention, probably the most significant international environmental justice treaty ever written. Some documents have been abridged, with the most salient sections provided and less important sections omitted. Editorial comments indicate the content of the deleted sections.

Documents

Civil Rights Act of 1964

The fundamental statement of the nation's position on civil rights is contained in the Civil Rights Act of 1964 (Public Law 88-352). This sweeping piece of legislation covered nearly every aspect of American public life, including institutionalized persons (Chapter I), public accommodations (Chapter II), public facilities (Chapter III), education (Chapter IV), and employment (Chapter VI). Three sections of the law provide the fundamental philosophy for the equal treatment of all Americans.

Sec. 1981. Equal Rights under the Law
(a) Statement of equal rights.   All persons within the jurisdiction of the United States shall have the same right in every State and Territory to make and enforce contracts, to sue, be parties, give evidence, and to the full and equal benefit of all laws and proceedings for the security of persons and property as is enjoyed by white citizens, and shall be subject to like punishment, pains, penalties, taxes, licenses, and exactions of every kind, and to no other.

Sec. 1982. Property Rights of Citizens
All citizens of the United States shall have the same right, in every State and Territory, as is enjoyed by white citizens thereof to inherit, purchase, lease, sell, hold, and convey real and personal property.

Sec. 2000d. Prohibition against exclusion from participation in, denial of benefits of, and discrimination under federally assisted programs on ground of race, color, or national origin

No person in the United States shall, on the ground of race, color, or national origin, be excluded from participation in, be denied the benefits of, or be subjected to discrimination under any program or activity receiving Federal financial assistance.

Source: U.S. Code Title 42, Chapter 21, Sections 1981 and 1982; and Title 42, Subchapter V, Section 2000d.

Fair Housing Act of 1968

Some experts believe that the Fair Housing Act of 1968 provides a strong justification for cases brought against companies responsible for the pollution of neighborhoods and communities. They argue that the act calls for all residents to receive equal treatment in the United States, but that the presence of disproportionate siting of polluting and hazardous waste sites violates that principle. Section 3601 lays out the general philosophy behind the act, and then section 3604 outlines some of the details of nondiscrimination required by the act. Probably the most important part of the following selection is section 3604(b).

Sec. 3601. Declaration of Policy

It is the policy of the United States to provide, within constitutional limitations, for fair housing throughout the United States.

Sec. 3604. Discrimination in the sale or rental of housing and other prohibited practices

As made applicable by section 3603 of this title and except as exempted by sections 3603(b) and 3607 of this title, it shall be unlawful—

(a) To refuse to sell or rent after the making of a bona fide offer, or to refuse to negotiate for the sale or rental of, or otherwise make unavailable or deny, a dwelling to any person because of race, color, religion, sex, familial status, or national origin.

(b) To discriminate against any person in the terms, conditions, or privileges of sale or rental of a dwelling, or in the provision of services or facilities in connection therewith, because of race, color, religion, sex, familial status, or national origin.

(c) To make, print, or publish, or cause to be made, printed, or published any notice, statement, or advertisement, with respect to the sale or rental of a dwelling that indicates any preference, limitation, or discrimination based on race, color, religion, sex, handicap, familial status, or national origin, or an intention to make any such preference, limitation, or discrimination.

(d) To represent to any person because of race, color, religion, sex, handicap, familial status, or national origin that any dwelling is not available for inspection, sale, or rental when such dwelling is in fact so available.

(e) For profit, to induce or attempt to induce any person to sell or rent any dwelling by representations regarding the entry or prospective entry into the neighborhood of a person or persons of a particular race, color, religion, sex, handicap, familial status, or national origin.

Source: U.S. Code Title 42, Chapter 45, Subchapter 1, Sections 3601 and 3604.

National Environmental Policy Act of 1969

The cornerstone of American environmental policy is the National Environmental Policy Act of 1969, now Chapter 55 of Title 42 of the U.S. Code. The act was signed by President Richard Nixon on January 1, 1970, and has been amended a number of times since then. Subchapter I of Chapter 55 lays out the nation's general policy regarding environmental issues. The sections below are generally used by proponents of environmental justice to provide the legal basis of their arguments in court cases and in the public forum.

Sec. 4331. Congressional declaration of national environmental policy

(a) The Congress, recognizing the profound impact of man's activity on the interrelations of all components of the natural environment, particularly the profound influences of population growth, high-density urbanization, industrial expansion, resource exploitation, and new and expanding technological advances and recognizing further the critical importance of restoring and maintaining environmental quality to the overall welfare and development of man, declares that it is the continuing policy of the Federal Government, in cooperation with State and local governments, and other concerned public and private organizations, to use all practicable means and measures, including financial and technical assistance, in a manner calculated to foster and promote the general welfare, to create and maintain conditions under which man and nature can exist in productive harmony, and fulfill the social, economic, and other requirements of present and future generations of Americans.

(b) In order to carry out the policy set forth in this chapter, it is the continuing responsibility of the Federal Government to use all practicable means, consistent with other essential considerations of national policy, to improve and coordinate Federal plans, functions, programs, and resources to the end that the Nation may—

(1) fulfill the responsibilities of each generation as trustee of the environment for succeeding generations;

(2) assure for all Americans safe, healthful, productive, and esthetically and culturally pleasing surroundings;

(3) attain the widest range of beneficial uses of the environment without degradation, risk to health or safety, or other undesirable and unintended consequences;

(4) preserve important historic, cultural, and natural aspects of our national heritage, and maintain, wherever possible, an environment which supports diversity and variety of individual choice;

(5) achieve a balance between population and resource use which will permit high standards of living and a wide sharing of life's amenities; and

(6) enhance the quality of renewable resources and approach the maximum attainable recycling of depletable resources.

(c) The Congress recognizes that each person should enjoy a healthful environment and that each person has a responsibility to contribute to the preservation and enhancement of the environment.

Sec. 4332. Cooperation of agencies; reports; availability of information; recommendations; international and national coordination of efforts

The Congress authorizes and directs that, to the fullest extent possible:

(1) the policies, regulations, and public laws of the United States shall be interpreted and administered in accordance with the policies set forth in this chapter, and

(2) all agencies of the Federal Government shall—

(A) utilize a systematic, interdisciplinary approach which will insure the integrated use of the natural and social sciences and the environmental design arts in planning and in decisionmaking which may have an impact on man's environment;

(B) identify and develop methods and procedures, in consultation with the Council on Environmental Quality established by subchapter II of this chapter, which will insure that presently unquantified environmental amenities and values may be given appropriate consideration in decisionmaking along with economic and technical considerations;

(C) include in every recommendation or report on proposals for legislation and other major Federal actions significantly affecting the quality of the human environment, a detailed statement by the responsible official on—

(i) the environmental impact of the proposed action,

(ii) any adverse environmental effects which cannot be avoided should the proposal be implemented,

(iii) alternatives to the proposed action,

(iv) the relationship between local short-term uses of man's environment and the maintenance and enhancement of long-term productivity, and

(v) any irreversible and irretrievable commitments of resources which would be involved in the proposed action should it be implemented.

Prior to making any detailed statement, the responsible Federal official shall consult with and obtain the comments of any Federal agency which has jurisdiction by law or special expertise with respect to any environmental impact involved. Copies of such statement and the comments and views of the appropriate Federal, State, and local agencies, which are authorized to develop and enforce environmental standards, shall be made available to the President, the Council on Environmental Quality and to the public as provided by section 552 of title 5, and shall accompany the proposal through the existing agency review processes;

(D) Any detailed statement required under subparagraph (C) after January 1, 1970, for any major Federal action funded under a program of grants to States shall not be deemed to be legally insufficient solely by reason of having been prepared by a State agency or official, if:

(i) the State agency or official has statewide jurisdiction and has the responsibility for such action,

(ii) the responsible Federal official furnishes guidance and participates in such preparation,

(iii) the responsible Federal official independently evaluates such statement prior to its approval and adoption, and

(iv) after January 1, 1976, the responsible Federal official provides early notification to, and solicits the views of, any other State or any Federal land management entity of any action or any alternative thereto which may have significant impacts upon such State or affected Federal land management entity and, if there is any disagreement on such impacts, prepares a written assessment of such impacts and views for incorporation into such detailed statement.

The procedures in this subparagraph shall not relieve the Federal official of his responsibilities for the scope, objectivity, and content of the entire statement or of any other responsibility under this chapter; and further, this subparagraph does not affect the legal sufficiency of statements prepared by State agencies with less than statewide jurisdiction.

(E) study, develop, and describe appropriate alternatives to recommended courses of action in any proposal which involves unresolved conflicts concerning alternative uses of available resources;

(F) recognize the worldwide and long-range character of environmental problems and, where consistent with the foreign policy of the United States, lend appropriate support to initiatives, resolutions, and programs designed to maximize international cooperation in anticipating and preventing a decline in the quality of mankind's world environment;

(G) make available to States, counties, municipalities, institutions, and individuals, advice and information useful in restoring, maintaining, and enhancing the quality of the environment;

(H) initiate and utilize ecological information in the planning and development of resource-oriented projects; and

(I) assist the Council on Environmental Quality established by subchapter II of this chapter.

Source: U.S. Code, Title 42, Chapter 55, Subchapter I, Sections 4331 and 4332.

Siting of Hazardous Waste Landfills and Their Correlation with Racial and Economic Status of Surrounding Communities (1983)

In 1982, Rep. James Florio (D-NJ) asked the General Accounting Office (GAO) to determine the correlation between hazardous waste landfill siting and racial and economic status. The GAO report is one of the first official governmental studies to clearly show the effects of environmental racism in the United States. The most important findings of the study are given below.

We found that:

—There are four offsite hazardous waste landfills in Region IV's eight States. Blacks make up the majority of the population in three of the four communities where the landfills are located. At least 26 percent of the population in all four communities have income below the poverty level and most of this population is Black.

—The determination of where a hazardous waste landfill will be located is currently a State responsibility. Federal regulations, effective in January 1983, require that selected sites meet minimal location standards. EPA has just begun its review process to determine if sites meet these standards.

—Federal legislation requires public participation in the hazardous waste landfill permit process except for the approval of disposal for polychlorinated biphenyls (PCBs), which are regulated under separate legislation that does not provide for public participation. Because of delays in issuing final regulations three of the four landfills in Region IV have not yet undergone the final permit process where public participation is required. The fourth is a PCB landfill and even though not subject to

TABLE 6.1
1980 Census Population, Income, and
Poverty Data for Census Areas Where Landfills Are Located

Landfill Number	Percent Black Population	Mean Family Income		Population below Poverty Level		
		All Races	Blacks	Number	Percent	Black
Chemical Waste Management, 626	90	$11,198	$10,752	265	42	100
SCA Services, 849	38	$16,371	$6,781	260	31	100
Industrial Chemical Co., 728	52	$18,996	$12,941	188	26	92
Warren County PCB Landfill, 804	66	$10,367	$9,285	256	32	90

Source: General Accounting Office. *Siting of Hazardous Waste Landfills and Their Correlation with Racial and Economic Status of Surrounding Communities,* Washington, DC: Report GAO/RCED-83-168, June 1, 1983.

Federal requirements, had undergone this process. Only one site in the Nation (in Region VI) has been granted a final hazardous waste landfill permit and had been subjected to the public participation process.

[The specific data on which the above conclusions were based are then reported on page 4 of the report.]

Source: General Accounting Office. *Siting of Hazardous Waste Landfills and Their Correlation with Racial and Economic Status of Surrounding Communities.* Gaithersburg, MD: General Accountability Office, June 1, 1983. http://archive.gao.gov/d48t13/121648.pdf.

Toxic Wastes and Race in the United States (1987)

In 1987, the Commission on Racial Justice of the United Church of Christ commissioned a study on the extent and character of environmental racism in the United States. That study is now regarded as one of the landmark pieces of research on environmental justice in U.S. history. As a result of its findings in that study, the commission made a number of recommendations to governmental and nongovernmental bodies as to how they should respond to the problems the commission discovered. Those recommendations are reprinted below.

—We urge the President of the United States to issue an executive order mandating federal agencies to consider the impact of current policies and regulations on racial and ethnic communities.

—We urge the formation of an Office of Hazardous Wastes and Racial and Ethnic Affairs by the U.S. Environmental Protection Agency. This office should insure that racial and ethnic concerns regarding hazardous wastes, such as the cleanup of uncontrolled sites, are adequately addressed. In addition, we urge the EPA to establish a National Advisory Council on Racial and Ethnic Concerns.

—We urge state governments to evaluate and make appropriate revisions in their criteria for the siting of new hazardous waste facilities to adequately take into account the racial and socioeconomic characteristics of potential host communities.

—We urge the U.S. Conference of Mayors, the National Conference of Black Mayors and the National League of Cities to convene a national conference to address these issues from a municipal perspective.

—We urge civil rights and political organizations to gear up voter registration campaigns as a means to further empower racial and ethnic communities to effectively respond to hazardous waste issues and to place hazardous wastes in racial and ethnic communities at the top of state and national legislative agendas.

—We urge local communities to initiate education and action programs around racial and ethnic concerns regarding hazardous wastes.

We also call for a series of additional actions. Of paramount importance are further epidemiological and demographic research and the provision of information on hazardous wastes to racial and ethnic communities.

Source: The Commission for Racial Justice (United Church of Christ). *Toxic Wastes and Race in the United States: A National Report on the Racial and Socio-Economic Characteristics of Communities with Hazardous Waste Sites*, pp. xv–xvi. Copyright 1987 United Church of Christ. All rights reserved. Used by permission.

Principles of Environmental Justice (1991)

The following position statement was adopted at the National People of Color Environmental Leadership Summit in Washington, D.C., October 24–27, 1991. The statement is still widely regarded as making up the environmental justice movement's defining principles.

Preamble
WE, THE PEOPLE OF COLOR, gathered together at this multinational People of Color Environmental Leadership Summit, to begin to build a

national and international movement of all peoples of color to fight the destruction and taking of our lands and communities, do hereby re-establish our spiritual interdependence to the sacredness of our Mother Earth; to respect and celebrate each of our cultures, languages and beliefs about the natural world and our roles in healing ourselves; to insure environmental justice; to promote economic alternatives which would contribute to the development of environmentally safe livelihoods; and, to secure our political, economic and cultural liberation that has been denied for over 500 years of colonization and oppression, resulting in the poisoning of our communities and land and the genocide of our peoples, do affirm and adopt these Principles of Environmental Justice:

1) Environmental Justice affirms the sacredness of Mother Earth, ecological unity and the interdependence of all species, and the right to be free from ecological destruction.

2) Environmental Justice demands that public policy be based on mutual respect and justice for all peoples, free from any form of discrimination or bias.

3) Environmental Justice mandates the right to ethical, balanced and responsible uses of land and renewable resources in the interest of a sustainable planet for humans and other living things.

4) Environmental Justice calls for universal protection from nuclear testing, extraction, production and disposal of toxic/hazardous wastes and poisons and nuclear testing that threaten the fundamental right to clean air, land, water, and food.

5) Environmental Justice affirms the fundamental right to political, economic, cultural and environmental self-determination of all peoples.

6) Environmental Justice demands the cessation of the production of all toxins, hazardous wastes, and radioactive materials, and that all past and current producers be held strictly accountable to the people for detoxification and the containment at the point of production.

7) Environmental Justice demands the right to participate as equal partners at every level of decision-making, including needs assessment, planning, implementation, enforcement and evaluation.

8) Environmental Justice affirms the right of all workers to a safe and healthy work environment without being forced to choose between an unsafe livelihood and unemployment. It also affirms the right of those who work at home to be free from environmental hazards.

9) Environmental Justice protects the right of victims of environmental injustice to receive full compensation and reparations for damages as well as quality health care.

10) Environmental Justice considers governmental acts of environmental injustice a violation of international law, the Universal Declaration on Human Rights, and the United Nations Convention on Genocide.

11) Environmental Justice must recognize a special legal and nat-ural relationship of Native Peoples to the U.S. government through treaties, agreements, compacts, and covenants affirming sovereignty and self-determination.

12) Environmental Justice affirms the need for urban and rural ecological policies to clean up and rebuild our cities and rural areas in balance with nature, honoring the cultural integrity of all our commu-nities, and provided fair access for all to the full range of resources.

13) Environmental Justice calls for the strict enforcement of princi-ples of informed consent, and a halt to the testing of experimental re-productive and medical procedures and vaccinations on people of color.

14) Environmental Justice opposes the destructive operations of multi-national corporations.

15) Environmental Justice opposes military occupation, repression and exploitation of lands, peoples and cultures, and other life forms.

16) Environmental Justice calls for the education of present and fu-ture generations which emphasizes social and environmental issues, based on our experience and an appreciation of our diverse cultural perspectives.

17) Environmental Justice requires that we, as individuals, make personal and consumer choices to consume as little of Mother Earth's resources and to produce as little waste as possible; and make the con-scious decision to challenge and reprioritize our lifestyles to insure the health of the natural world for present and future generations.

Source: Environmental Justice Resource Center. http://www.ejrc.cau .edu/princej.html. Accessed May 14, 2008.

Environmental Equity: Reducing Risks for All Communities (1992)

In 1991, Environmental Protection Agency administrator William K. Reilly appointed a commission to study environmental inequities in the United States based on race, income status, and other factors. The com-mission released its report on July 22, 1992, the main features of which are quoted below.

Summary of Findings

- There are clear differences between racial groups in terms of disease and death rates. There are also limited data to ex-plain the environmental contribution to these differences. In fact, there is a general lack of data on environmental health ef-fects by race and income. For diseases that are known to have

environmental causes, data are not typically dis-aggregated by
race and socioeconomic group. The notable exception is lead
poisoning: A significantly higher percentage of Black children
compared to White children have unacceptably high blood lead
levels.

- Racial minority and low-income populations experience higher
  than average exposures to selected air pollutants, hazardous
  waste facilities, contaminated fish and agricultural pesticides in
  the workplace. Exposure does not always result in an immediate
  or acute health effect. High exposures, and the possibility of
  chronic effects, are nevertheless a clear cause for health concerns.

- Environmental and health data are not routinely collected and
  analyzed by income and race. Nor are data routinely collected
  on health risks posed by multiple industrial facilities, cumulative and synergistic effects, or multiple and different pathways
  of exposure. Risk assessment and risk management procedures
  are not in themselves biased against certain income or racial
  groups. However, risk assessment and risk management procedures can be improved to better take into account equity
  considerations.

- Great opportunities exist for EPA and other government agencies to improve communication about environmental problems
  with members of low-income and racial minority groups. The
  language, format and distribution of written materials, media
  relations, and efforts in two-way communication all can be improved. In addition, EPA can broaden the spectrum of groups
  with which it interacts.

- Since they have broad contact with affected communities, EPA's
  program and regional offices are well suited to address equity
  concerns. The potential exists for effective action by such offices
  to address disproportionate risks. These offices currently vary
  considerably in terms of how they address environmental equity issues. Case studies of EPA program and regional offices reveal that opportunities exist for addressing environmental
  equity issues and that there is a need for environmental equity
  awareness training. A number of EPA regional offices have initiated projects to address high risks in racial minority and low-income communities.

- Native Americans are a unique racial group that has a special
  relationship with the federal government and distinct environmental problems. Tribes often lack the physical infrastructure,
  institutions, trained personnel and resources necessary to protect their members.

Summary of Recommendations

Although large gaps in data exist, the Workgroup believes that enough is known with sufficient certainty to make several recommendations to the Agency. These recommendations are also applicable to other public and private groups engaged in environmental protection activities. The job of achieving environmental equity is shared by everyone.

- EPA should increase the priority that it gives to issues of environmental equity.

- EPA should establish and maintain information which provides an objective basis for assessment of risks by income and race, beginning with the development of a research and data collection plan.

- EPA should incorporate considerations of environmental equity into the risk assessment process. It should revise its risk assessment procedures to ensure, where practical and relevant, better characterization of risk across populations, communities or geographic areas. These revisions could be useful in determining whether there are any population groups at disproportionately high risk.

- EPA should identify and target opportunities to reduce high concentrations of risk to specific population groups, employing approaches developed for geographic targeting.

- EPA should, where appropriate, assess and consider the distribution of projected risk reduction in major rulemakings and Agency initiatives.

- EPA should selectively review and revise its permit, grant, monitoring and enforcement procedures to address high concentrations of risk in racial minority and low-income communities. Since state and local governments have primary authority for many environmental programs, EPA should emphasize its concerns about environmental equity to them.

- EPA should expand and improve the level and forms with which it communicates with racial minority and low-income communities and should increase efforts to involve them in environmental policy-making.

- EPA should establish mechanisms, including a center of staff support, to ensure that environmental equity concerns are incorporated in its long-term planning and operations.

Source: "Release of Environmental Equity Report." http://www. epa.gov/history/topics/justice/01.htm. Accessed May 14, 2008.

Executive Order 12898 (President Clinton)— Federal Actions to Address Environmental Justice in Minority Populations and Low-Income Populations (1994)

One of the most important documents in the history of the environmental justice movement in the United States is Executive Order 12898, issued by President Bill Clinton on February 11, 1994. In this order, Clinton established a national policy for the role of environmental justice issues in the federal government and provided a structure by which this policy was to be implemented.

February 11, 1994
By the authority vested in me as President by the Constitution and the laws of the United States of America, it is hereby ordered as follows:

Section 1-1. Implementation

1-101. Agency Responsibilities.   To the greatest extent practicable and permitted by law, and consistent with the principles set forth in the report on the National Performance Review, each Federal agency shall make achieving environmental justice part of its mission by identifying and addressing, as appropriate, disproportionately high and adverse human health or environmental effects of its programs, policies, and activities on minority populations and low-income populations in the United States and its territories and possessions, the District of Columbia, the Commonwealth of Puerto Rico, and the Commonwealth of the Mariana Islands.

1-102. Creation of an Interagency Working Group on Environmental Justice.   (a) Within 3 months of the date of this order, the Administrator of the Environmental Protection Agency ("Administrator") or the Administrator's designee shall convene an interagency Federal Working Group on Environmental Justice ("Working Group"). The Working Group shall comprise the heads of the following executive agencies and offices, or their designees: (a) Department of Defense; (b) Department of Health and Human Services; (c) Department of Housing and Urban Development; (d) Department of Labor; (e) Department of Agriculture; (f) Department of Transportation; (g) Department of Justice; (h) Department of the Interior; (i) Department of Commerce; (j) Department of Energy; (k) Environmental Protection Agency; (l) Office of Management and Budget; (m) Office of Science and Technology Policy;

(n) Office of the Deputy Assistant to the President for Environmental Policy; (o) Office of the Assistant to the President for Domestic Policy; (p) National Economic Council; (q) Council of Economic Advisers; and (r) such other Government officials as the President may designate. The Working Group shall report to the President through the Deputy Assistant to the President for Environmental Policy and the Assistant to the President for Domestic Policy.

 (b) The Working Group shall:

1. provide guidance to Federal agencies on criteria for identifying disproportionately high and adverse human health or environmental effects on minority populations and low-income populations;

2. coordinate with, provide guidance to, and serve as a clearinghouse for, each Federal agency as it develops an environmental justice strategy as required by section 1–103 of this order, in order to ensure that the administration, interpretation and enforcement of programs, activities and policies are undertaken in a consistent manner;

3. assist in coordinating research by, and stimulating cooperation among, the Environmental Protection Agency, the Department of Health and Human Services, the Department of Housing and Urban Development, and other agencies conducting research or other activities in accordance with section 3–3 of this order;

4. assust in coordinating data collection, required by this order;

5. examine existing data and studies on environmental justice;

6. hold public meetings as required in section 5–502(d) of this order; and

7. develop interagency model projects on environmental justice that evidence cooperation among Federal agencies.

1-103. Development of Agency Strategies. (a) Except as provided in section 6–605 of this order, each Federal agency shall develop an agency-wide environmental justice strategy, as set forth in subsections (b)–(e) of this section that identifies and addresses disproportionately high and adverse human health or environmental effects of its programs, policies, and activities on minority populations and low-income populations. The environmental justice strategy shall list programs, policies, planning and public participation processes, enforcement, and/or rulemakings related to human health or the environment that should be revised to, at a minimum: (1) promote enforcement of all health and environmental statutes in areas with minority populations and low-income populations; (2) ensure greater public participation; (3) improve research and data collection relating to the health of and environment of minority populations and low-income populations; and (4) identify differential patterns of consumption of natural resources among minority populations and low-income populations. In addition, the environmental justice strategy shall include, where appropriate, a timetable for undertaking identified revisions and consideration of economic and social implications of the revisions.

[Sections (b) through (g) and 1–104 provide "housekeeping" rules for the conduct of these responsibilities.]

Section 2-2. Federal Agency Responsibilities for Federal Programs
Each Federal agency shall conduct its programs, policies, and activities that substantially affect human health or the environment, in a manner that ensures that such programs, policies, and activities do not have the effect of excluding persons (including populations) from participation in, denying persons (including populations) the benefits of, or subjecting persons (including populations) to discrimination under, such programs, policies, and activities, because of their race, color, or national origin.

Section 3-3. Research, Data Collection, and Analysis
[This section directs all agencies to collect and analyze data and carry out research on environmental risks faced by minority and low-income populations and workers who may be exposed to "substantial environmental risks."]

Section 4-4. Subsistence Consumption of Fish and Wildlife
[This section directs agencies to conduct research on the consumption of fish and wildlife as a result of exposure to hazardous materials and to provide nutritional guidance to populations that may be at risk because of their consumption of such foods.]

Section 5-5. Public Participation and Access to Information
[This section clarifies the right of the general public to provide input on environmental justice issues to federal agencies and the manner by which that information is to be utilized by those agencies.]

Section 6-6. General Provisions
[This section provides additional "housekeeping" instructions about the relationships of this executive order to other laws and regulations, special cases, costs, and judicial review.]

Source: Executive Order no. 12898. *Federal Register* 59, no. 32 (February 16, 1994): 7629–7633.

Not in My Backyard: Executive Order 12898 and Title VI as Tools for Achieving Environmental Justice (2003)

In 2003, the U.S. Commission on Civil Rights undertook a study to determine how well four government agencies—the Environmental Protection Agency, the U.S. Department of the Interior, the U.S. Department of Housing and Urban Development, and the U.S. Department of Transportation—have implemented Executive Order 12898 and Title VI of the 1964 Civil Rights Act. The commission concluded that these agencies had not made much progress in this regard. In its letter of transmittal to the president and the Congress, the commission made the following observation.

[S]ignificant problems and shortcomings remain. Federal agencies still have neither fully incorporated environmental justice into their core missions nor established accountability and performance outcomes for programs and activities. Moreover, a commitment to environmental justice is often lacking in agency leadership, communities are not yet full participants in environmental decision-making, and there is still inadequate scientific and technical literature on the relationship between environmental pollutants and human health status. Although poor communities and communities of color are becoming more skilled at using Title VI administrative processes to seek recourse and remedies, agencies seldom, if ever, revoke a permit or withhold money from the recipients of federal funding for violating Title VI. Strong administrative enforcement of Title VI is required in light of court decisions limiting access to judicial recourse and remedies under Title VI. Uncertainty about the use and effectiveness of Title VI in protecting the poor and communities of color is created by the absence of final investigative and recipient guidance by EPA. The agency was moving toward finalizing its Title VI guidance at the time the Commission report was drafted, and we look forward to its release. The other agencies, unlike EPA, lacked any comprehensive Title VI investigation and recipient guidance.

[Some of the conclusions offered by the commission are as follows:]

- A renewed effort by federal agencies to collect, analyze, and maintain data on risks and exposures be undertaken.
- Formal guidance on assessing cumulative risk should be created by federal agencies that considers the roles of social, economic, and behavioral factors when assessing risk.
- Guidance should include a presumption of adverse health risks when populations are exposed to multiple hazards from multiple sources.

- Federal agencies should disaggregate data on risks and exposures by race, ethnicity, gender, age, income, and geographic location if communities are to have the tools they need to defend environmental and human health and if agencies are to fulfill their obligations under Executive Order 12898 and Title VI.
- Federal agencies should require state and local zoning and land-use authorities, as a condition for receiving and continuing to receive federal funding, to incorporate and implement the principles of environmental justice into their zoning and land-use policies.
- The funding scheme for the Superfund program should be reviewed by Congress to ensure that the program is effectively funded and administered.

Source: U.S. Commission on Civil Rights. *Not in My Backyard: Executive Order 12898 and Title VI as Tools for Achieving Environmental Justice.* Washington, DC: U.S. Commission on Civil Rights, October 2003, pp. iii, 27–28.

2003–2008 EPA Strategic Plan: Direction for the Future (2003)

In September 2003, the U.S. Environmental Protection Agency published a five-year plan for the agency that focused on five major goals: clean air and addressing global climate change, clean and safe water, land preservation and restoration, healthy communities and ecosystems, and compliance and environmental stewardship. Each goal included some discussion of environmental justice, but by far the greatest emphasis occurred in Goal 4: Healthy Communities and Ecosystems. The section on environmental justice in that part of the report included the following discussion.

Restoring Healthy Communities: Environmental Justice

EPA remains committed to environmental justice for all people, regardless of race, color, national origin, or income, in accordance with Executive Order 12898, "Federal Actions to Address Environmental Justice in Minority Populations and Low-Income Populations." Recognizing that minority and/or low-income communities may be disproportionately exposed to environmental hazards and risks, we will work to protect these and other affected communities. Environmental justice means not only protecting human health and the environment for everyone, but also ensuring that all people are treated fairly and are given opportunities to participate meaningfully in developing, implementing, and enforcing environmental laws, regulations, and policies.

EPA is establishing measurable environmental justice commitments for eight national priorities: reducing asthma attacks, reducing exposure to air toxics, increasing compliance with regulations, reducing incidence of elevated blood lead levels, ensuring that fish and shellfish are safe to eat, ensuring that water is safe to drink, revitalizing brownfields and contaminated sites, and using collaborative problem-solving to address environmental and public health concerns. We will promote environmental justice in all aspects of our work by training staff; providing guidance, online tools, and other resources; sharing information about successful strategies; and enhancing staff skills in working with community-based organizations. We will continue to use dispute resolution, facilitation, listening sessions, and other consensus-building techniques and to convene stakeholders to address environmental and public health issues.

Assessing and Cleaning Up Brownfields

Brownfields are real properties where expansion, redevelopment, or reuse may be complicated by the presence or potential presence of hazardous substances, pollutants, or contaminants. Assessing brownfields can help communities understand the risks these properties pose and provides the information needed to undertake cleanup and reuse. Cleaning up and reinvesting in these properties may increase local tax bases, facilitate job growth, utilize existing infrastructure, take development pressures off undeveloped land, and improve and protect the environment. EPA will continue to award competitive grants to assess and clean up brownfields and to provide job training opportunities within affected communities.

Awards are based on a number of factors including how well the project reduces threats to human health and the environment, and creates and/or preserves greenspace. In addition, the Brownfields Revitalization Act requires us to consider "the extent to which the grant would address or facilitate the identification and reduction of threats to the health or welfare of minority or low-income communities, or other sensitive populations," underlining our commitment to environmental justice. Our Brownfields Program is also developing a methodology to assess the relationship between EPA-funded brownfields projects and the sensitive, socio-economically disadvantaged communities that they serve. EPA will use this methodology to improve how the Brownfields Program incorporates environmental justice concerns into its operations.

We will continue to provide funds to state and tribal governments to establish and enhance response programs that oversee the majority of brownfields assessments and cleanups. These programs provide technical oversight and assist property owners; create inventories of brownfields sites; and develop policies, regulations, and ordinances. Funding can also be used to conduct assessment and cleanup activities at brownfields properties. EPA funding is often critical for operating these response programs, particularly for tribal governments.

We will also continue to provide outreach and technical assistance to communities confronting brownfields and perform targeted assessments at sites where stakeholders are seeking federal assistance to identify the extent of contamination. Through the Brownfields and Land Revitalization Technology Support Center, we will help streamline site investigations and cleanup processes, identify technology options, evaluate contractor capabilities and recommendations, and explain complex technologies to communities. Technical tools such as Triad and SMARTe can aid communities' brownfields efforts. EPA will continue to sponsor brownfields workshops and educational events that provide forums for sharing ideas, lessons learned, and best practices.

Reducing Transboundary Threats along the U.S.–Mexico Border

The U.S.-Mexico Border 2012 Program, a joint effort between the governments of the United States and Mexico, works with the 10 border states and border communities to reduce transboundary threats to improve the region's environmental and ecosystem health. As part of our continuing commitment to environmental justice, EPA is working with some disadvantaged border communities to improve water quality in both the United States and Mexico. For decades, raw sewage posed a significant public health and environmental threat to U.S. and Mexican communities. Inadequate water and sewage treatment cause border residents to suffer disproportionately from hepatitis A and other waterborne diseases. EPA assists communities in the U.S.-Mexico border region to increase the number of homes with access to safe drinking water and basic sanitation. As this infrastructure comes on line, discharges of raw sewage will be reduced and surface water quality will improve. Restoration of surface water quality on 10 impaired transboundary waters is an EPA priority. EPA also will address health and environmental risks presented by abandoned tires and hazardous waste. Piles of waste tires breed mosquitoes and other disease-carrying organisms, and they are prone to fires that are difficult to extinguish. Contaminated hazardous waste sites pose acute and long-term risks from metal poisoning. We will address key sites on the border, laying the foundation for future remediation efforts.

Source: 2003–2008 EPA Strategic Plan: Direction for the Future. Washington, DC: U.S. Environmental Protection Agency, September 30, 2003, pp. 91–93.

Toolkit for Assessing Potential Allegations of Environmental Injustice (2004)

The ultimate federal authority for dealing with claims of environmental inequities in the United States is the EPA. The agency has developed the "Toolkit for Assessing Potential Allegations of Environmental Injustice" for use when possible instances of environmental injustice are brought before it. The Toolkit outlines four stages for analysis of an alleged case of environmental injustice: (1) problem formulation, (2) data collection, (3) assessment of the potential for adverse environmental and human health effects, and (4) assessment of the potential for disproportionately high and adverse effects. A key part of this scheme is stage 2, data collection. To aid agency employees in collecting data relevant to a claim of environmental inequity, the EPA has developed a detailed list of relevant factors. They are as follows:

(1) Identification of Environmental Sources of Stress and Likelihood of Exposure.

Information is collected to determine the sources of environmental pressures or stress in the affected area that are different from or additional to those in the reference community. This includes an assessment of existing or likely future additional pressures or sources of stress and their proximity to the community (compared with the reference community), that may result from a particular decision or activity. Specific types of data to be considered include:

- Sources of stressors placed on the community
  - Number of environmentally regulated facilities within a community
  - Length of time regulated facilities have operated within a community
  - Number of current and past permit exceedances by regulated facilities
  - Number or extent of non-point sources of pollution
  - Noise levels
- Potential exposure to stressors
  - Proximity of regulated facilities to the majority of the community's population
  - Proximity to multiple contaminant sources
  - Potential or actual cumulative exposure across multiple locations
  - Potential or actual exposure to multiple stressors
  - Number of biomarkers of exposure that are evident
- Environmental conditions resulting from stressors
  - "Quality" of the air, water, and other environmental media
  - Density of contaminants in biota (living organisms)

- Environmental vulnerability
  - Climate
  - Geomorphic features
  - Hydrogeomorphic features
  - Presence of ecologically sensitive areas

(2) Collection of Data on Affected Area and Reference Community. Health-related, social, and economic data on the affected area and on the potential reference communities are collected. This information will be used to compare impacts on the affected area in comparison with the reference community. Generally speaking, the geographic boundaries for a potentially affected area will be defined by the problem (e.g., defining an impact zone around a hazardous waste site or permitted facility) or by the community itself.

Health Indicators provide information on the general health of the community's residents and their ability to cope with environmental stresses. Specific types of data that would be considered include:

- Existing health conditions
  - Percent of infant mortality within the community (per 1000 births)
  - Average birth weight
  - Adult mortality
  - Life expectancy at birth
- Health impacts from environmental stressors
  - Number of illnesses attributable to chemical contaminants— if contaminant stressors are in question.
  - Number of diseases attributable to pathogens—if pathogenic stressors are in question.

Social Indicators include data on the distribution of certain population characteristics (e.g., race, color, or ethnicity). Specific types of data that would be considered include:

- General demographics
  - Percent of population (that have various ethnic and national origins)
  - Population density, including the distribution of urban and rural populations
  - Percent of the population that is Native American
  - Distribution of languages spoken in population
  - Percent of the population that is literate in English or other languages
- Vulnerability to exposure
  - Percent of community with access to public transportation and services

- Percent of community with access to health care facilities
- Percent of community that uses regulated (cigarettes, alcohol) and unregulated (drugs) substances
- Percent of community with access to alternative sources of drinking water
- Percent of community with sewage treatment
- Percent of community that relies on local food sources
- Government response actions
  - Expenditure/investment on providing access to environmental information (as a percent of total community budget)
  - Expenditure/investment on environmental education and training (as a percent of total community budget)
  - Number and frequency of public meetings on proposed actions and policy decisions
  - Number of different types of materials distributed
  - Percent of households that received distributed materials
  - Number of documents available in the various languages associated with a community
- Community participation
  - Community identification
  - Cultural dynamics
  - Quality of public participation of community residents
  - Number of community residents participating in non-governmental organizations
  - Number of community members participating in the decision-making process

Economic Indicators reveal trends about the community's economic wellbeing. Specific types of data that would be considered include:

- Economic information
  - Unemployment rate
  - Income level and distribution
  - Percent of homeowners in a community or the percent of renters in a community
  - Percent of community residents with employment in pollution-generating industrial facilities or services
  - Number of brownfields in the community
  - Reliance on natural resources for the community's economic base (as a percent of total community budget)

Source: Desk Reference to the Toolkit for Assessing Potential Allegations of Environmental Injustice. Washington, DC: Environmental Protection Agency, 2006, pp. 2–5. http://www.epa.gov/compliance/resources/policies/ej/ej-toolkit-desk-ref.pdf.

Evaluation Report: EPA Needs to Consistently Implement the Intent of the Executive Order on Environmental Justice (2004)

In late 2003, the Office of Inspector General of the EPA initiated a study to determine the extent to which the EPA had implemented Executive Order 12898 on Environmental Justice, issued by President Bill Clinton a decade earlier. As the title of the report suggests, the inspector general found that the agency had failed in a number of ways to carry out the spirit and the letter of the executive order. The report's general conclusions are summarized below. Note that the EPA was required to respond to the report; that response is appended as a part of this report.

Executive Summary

Purpose

In 1994, President Clinton issued Executive Order 12898, "Federal Action to Address Environmental Justice in Minority Populations and Low-Income Populations," to ensure such populations are not subjected to a disproportionately high level of environmental risk. The overall objective of this evaluation was to determine how the U.S. Environmental Protection Agency (EPA) is integrating environmental justice into its day-to-day operations. Specifically, we sought to answer the following questions:

- How has the Agency implemented Executive Order 12898 and integrated its concepts into EPA's regional and program offices?
- How are environmental justice areas defined at the regional levels and what is the impact?

Results in Brief

EPA has not fully implemented Executive Order 12898 nor consistently integrated environmental justice into its day-to-day operations. EPA has not identified minority and low-income, nor identified populations addressed in the Executive Order, and has neither defined nor developed criteria for determining disproportionately impacted *[definitional footnote omitted here].* Moreover, in 2001, the Agency restated its commitment to environmental justice in a manner that does not emphasize minority and low-income populations, the intent of the Executive Order.

Although the Agency has been actively involved in implementing Executive Order 12898 for 10 years, it has not developed a clear vision or a comprehensive strategic plan, and has not established values, goals, expectations, and performance measurements. We did note that the Agency made an attempt to issue an environmental justice toolkit; endorsed environmental justice training; and required that all regional and programmatic offices submit "Action Plans" to develop some accountability for environmental justice integration.

In the absence of environmental justice definitions, criteria, or standards from the Agency, many regional and program offices have taken steps, individually, to implement environmental justice policies. This has resulted in inconsistent approaches by the regional offices. Thus, the implementation of environmental justice actions is dependent not only on minority and income status but on the EPA region in which the person resides. Our comparison of how environmental justice protocols used by three different regions would apply to the same city showed a wide disparity in protected populations.

We believe the Agency is bound by the requirements of Executive Order 12898 and does not have the authority to reinterpret the order. The Acting Deputy Administrator needs to reaffirm that the Executive Order 12898 applies specifically to minority and low-income populations that are disproportionately impacted. After 10 years, there is an urgent need for the Agency to standardize environmental justice definitions, goals, and measurements for the consistent implementation and integration of environmental justice at EPA.

Recommendations

We recommended that the Acting Deputy Administrator issue a memorandum reaffirming that Executive Order 12898 is an Agency priority and that minority and low-income populations disproportionately impacted will be the beneficiaries of this Executive Order. Additionally, EPA should establish specific time frames for the development of definitions, goals, and measurements. Furthermore, we recommended that EPA develop and articulate a clear vision on the Agency's approach to environmental justice. We also recommended that EPA develop a comprehensive strategic plan, ensure appropriate training is provided, clearly define the mission of the Office of Environmental Justice, determine if adequate resources are being applied to environmental justice, and develop a systematic approach to gathering information related to environmental justice.

Agency Comments and OIG Evaluation

In the response to our draft report, the Agency disagreed with the central premise that Executive Order 12898 requires the Agency to identify and address the environmental effects of its programs on minority and low-income populations. The Agency believes the Executive Order "instructs the Agency to identify and address the disproportionately high and adverse human health or environmental effects of it [*sic*] programs, policies, and activities." The Agency does not take into account the inclusion of the minority and low-income populations, and indicated it is attempting to provide environmental justice for everyone. While providing adequate environmental justice to the entire population is commendable, doing so had already been EPA's mission prior to implementation of the Executive Order; we do not believe the intent of the Executive Order was simply to reiterate that mission.

A summary of the Agency's response and our evaluation is included at the end of each chapter. The Agency's complete response and our evaluation of that response are included in Appendices D and E, respectively.

Source: Office of Inspector General. *EPA Needs to Consistently Implement the Intent of the Executive Order on Environmental Justice.* Report No. 2004-P-00007. Washington, DC: U.S. Environmental Protection Agency, March 1, 2004, pp. i–ii.

Redefining Rights in America: The Civil Rights Record of the George W. Bush Administration, 2001–2004 (2004)

In 2004, the U.S. Commission on Civil Rights undertook a comprehensive review of the civil rights record of the first administration of President George W. Bush. The review covered subjects such as judicial nominations, federal appointments, the federal workforce, voting rights, fair housing, affirmative action, racial profiling, and environmental justice. The report is no longer easily available from the commission, but a draft of the final report is available at a number of sites online. The section of the report dealing with environmental justice is reprinted below. Numerous footnotes and other citations are omitted from the selection.

Environmental Justice

Researchers have asserted that "there are times when environmental problems raise important civil rights questions." Civil rights violations occur when certain communities, especially black, Hispanic, and Native American, are inequitably burdened by environmental ills. There are many historical and present-day examples of environmental injustice. Among them:

- Public and private initiatives have targeted minority communities for locating toxic facilities, such as incinerators, oil refineries, power plants, landfills, and diesel bus stations;

- Land-use policies and unhealthy and hazardous conditions have uprooted working class minority communities from neighborhoods; and
- Low-income people and minorities have been excluded from decisionmaking regarding environmental policies, programs, and permits that affected them.

Myriad reasons account for the fact that many hazardous waste sites are located in minority communities, ranging from discrimination to economic motivation. Regardless, minority communities bear a disproportionate share of the burden of these sites, including exposure to hazardous conditions and life-threatening chemicals, such as pesticides and cancer-causing solvents, a circumstance that has led people of color to increasingly seek justice from discriminatory environmental policies over the past three decades.

Environmental justice seeks the equitable treatment of all racial and income groups and cultures in the development, execution, and enforcement of environmental laws, rules, and policies, including their meaningful participation in the decisionmaking processes of the government. Further, environmental justice provides a framework for minorities and low-income populations to identify the continually reinforced political and economic assumptions underlying environmental injustices.

A History of Environmental Justice Policies

In 1970, to better address pollution concerns, President Nixon reorganized environmental functions of federal agencies and created the Environmental Protection Agency (EPA). As a federal grant-making agency, EPA has an obligation to ensure compliance with Title VI of the Civil Rights Act of 1964, which prohibits discrimination based on race, color, or national origin in programs and activities receiving federal financial assistance and requires federal agencies to regulate against discriminatory practices. The law protects against both intentional discrimination and actions that have a disparate impact on minorities, although the Supreme Court has determined that individuals can only pursue cases alleging the former. They must rely on federal agencies to bring disparate impact environmental justice lawsuits against entities that receive federal funding. The statute and Title VI regulations provide EPA with the authority to promote environmental justice policies.

In the 1970s, mounting evidence of discrimination prompted the birth of the environmental justice movement. In 1979, the residents of an East Houston, Texas, community alleged that the decision to place a garbage dump in their neighborhood was racially motivated and in violation of their civil rights. Although the court found that the placement of the dump would irreparably harm the community, it was unable to establish whether intentional discrimination had occurred or whether the site's placement reflected a discriminatory pattern. The case, however, launched the use of courts as a tool and highlighted the need to collect data and make it available to communities challenging environmental decisions.

In 1982, a citizen protest against a dump containing highly toxic waste, which the state of North Carolina was forcing on one of the poorest counties in the state with a population that was more than 84 percent black, captured national attention. As a result, a congressional delegate asked the U.S. General Accounting Office to conduct a study of hazardous waste landfill sites in the region. The study located four hazardous commercial waste landfills—three were in predominantly black communities and the fourth was in a low-income neighborhood.

As evidence mounted, the need for strong enforcement became clear. EPA issued Title VI regulations in 1972. In 1984, EPA strengthened the regulations to grant the administrator authority to refuse, delay, or discontinue funding to any program recipient operating in a discriminatory manner. The regulations, however, were not enforced against state and local funding recipients until 1993.

The documented patterns of discriminatory dumping and the lack of environmental justice enforcement precipitated President Clinton's 1995 [sic] issuance of Executive Order 12898, "Federal Actions to Address Environmental Justice in Minority Populations and Low-Income Populations." The order directs federal agencies to develop agencywide environmental justice strategies and to review their programs, policies, and activities that have a negative environmental impact on minority and low-income communities. Federal agencies are required to create accountability standards and measures to evaluate the goals of the executive order, with EPA the lead agency.

Integration of Environmental Equity in EPA Policy

To promote Executive Order 12898, in August 2001 President Bush asked then EPA Administrator Christine Todd Whitman to issue a memorandum affirming the administration's commitment to environmental justice, including its integration into all programs, policies, and activities. EPA headquarters and regional offices responded to the memorandum by developing action plans for obtaining measurable justice outcomes. The plans mandate that information be shared between program offices; establish data collection, management, and evaluation standards; and require that information be shared with and input accepted from external stakeholders. Program and regional offices began implementing the plans in 2003.

In the summer of 2003, EPA also launched a $1.5 million Environmental Justice Collaborative Problem Solving Grant Program for 15 nonprofit community-based organizations, the goal of which is to assist community organizations in finding viable solutions. Among the beneficiaries are the Coalition for West Oakland Revitalization, which is addressing the air quality problem in one of the poorest neighborhoods in the Bay Area of California, and the Pioneer Valley Project, Inc., which seeks to resolve health and related economic problems of Vietnamese women employed in nail salons in Springfield, Massachusetts, and surrounding areas caused by exposure to hazardous chemicals. Despite these initiatives, the National Academy of Public Administration (NAPA) recommended that EPA set clear expectations for producing results and that the agency establish accountability standards. According to NAPA, the administrator's August 2001 memorandum was an example of how the agency espouses strong language and expectations, but fails to provide specific agencywide measures of accountability. The study identified several reasons for the deficiency, including EPA's failure to establish goals for specific outcomes or adopt methods for measuring progress. Witnesses testifying before the Commission at a 2002 hearing on environmental justice stated that similar problems are found at other agencies. Since EPA does not identify specific outcomes nor measure progress, it is difficult to assess whether progress has been made toward its goals.

Furthermore, EPA's inspector general has determined that the former Bush administrator's memorandum changed the focus of the environmental justice program by deemphasizing minority and low-income populations. The inspector general also found that despite a decade of active pursuit, Executive Order 12898 is still not part of EPA's core mission. President Bush did not consider implementation of the order a primary goal. The inspector general notes that EPA has not developed a clear vision or a comprehensive strategic plan, nor established values, goals, expectations, and performance measurements regarding the order. Further, EPA has not provided regional or program offices with standards for what constitutes a minority or low-income community, or defined the term "disproportionately" as it relates to environmental justice. If EPA does not identify parameters for environmental justice, it will not be able to comply with Executive Order 12898.

Environmental Justice Enforcement and Guidance

In May 2001, Administrator Whitman formed a task force to resolve 66 open Title VI complaints that had accumulated at EPA between 1998 and 2001. Whitman assembled the task force to take advantage of a congressional decision to allow EPA to use its 2002 appropriations to investigate and resolve Title VI complaints. By February 8, 2002, Whitman had reduced the complaint backlog to 41. Of the 41 remaining cases, 34 were identified as acceptable for investigation. Of the 23 complaints filed since February 2002, two were accepted for investigation and 13 were under review as of May 2004. Although EPA was commended for reducing the existing backlog and accepting complaints for investigation, environmental justice experts testified before the Commission that some had been improperly handled. For example, a complaint against the Michigan Department of Environmental Quality concerning its issuance of permits for a proposed steel mill was rejected without an investigation into its disparate impact claim. According to one researcher, EPA's neglect of the disparate impact claim raises concerns that the task force's objective to reduce the backlog supercedes its obligation to appropriately investigate complaints.

These concerns merit attention because EPA receives the bulk of Title VI environmental grievances and has taken the lead in providing guidance to environmental stakeholders, advocates, and legal scholars. Adding to uncertainties over EPA's complaint processing is its failure to issue final Title VI guidance defining: (1) what constitutes disparate impact; (2) when complaints can be filed; (3) how long complaints take to process; (4) how communities are given information about participation in decisionmaking; and (5) how the interests of industry and communities can be balanced. Although EPA has issued various Title VI interim guidance on these issues, final guidance is needed to inform communities continually exposed to environmental pollutants of the important elements of an adverse disparate impact violation. Furthermore, the delay has left state and local regulators, among others, unable to determine when an industrial facility may be violating Title VI.

Defining Environmental Standards and Identifying Hazards

The Bush administration has undertaken several actions that undermine environmental justice. In so doing, it seems to believe that minority and low-income populations are not disproportionately affected by environmental pollutants. For example, a 2003 EPA report failed to embrace the notion that poor and disadvantaged populations reside in areas with higher concentrations of pollutants, or that the distribution of environmental burdens is based on race, income, and political power. According to researchers, this claim was a clear reversal of EPA's historical stand that minority and low-income communities are overburdened with environmental pollutants and an apparent retreat from Executive Order 12898. EPA and other federal agencies typically do not cooperate with health policy experts and affected minority and low-income communities to eliminate or reduce environmental pollutants.

Furthermore, under President Bush, the EPA has yet to develop a standard for assessing the cumulative impact of environmental hazards. Cumulative impact is the "threat to public health caused by the exposure to the sum total of releases" of these hazards. EPA offices, including the Office for Civil Rights, have published guidance on measuring the risks. The agency has not, however, developed a cumulative impact standard, indicating that the effort to establish a causal relationship between pollutants and health problems is difficult. EPA's delay prevents the government from eliminating the hazardous environmental conditions foisted upon minority communities. For example, one researcher described a mostly minority community in a region between Baton Rouge and New Orleans, Louisiana, where the cumulative environmental hazards produced by local industry far exceed the U.S. average. According to this researcher, environmental pollutants are so numerous that the region is known as "Cancer Alley."

Fostering Public Participation

The administration has also failed to increase the participation of affected minority and low-income communities in information gathering and dissemination and decisionmaking processes. Meaningful public participation by affected communities in the decisionmaking process is one of the cornerstones of environmental justice. Furthermore, community input is integral to planning, monitoring, problem solving, implementation, and evaluation of environmental policies and practices. Yet, EPA has not conducted public meetings to explain Executive Order 12898, its other environmental justice policies, or how affected communities could participate in decisionmaking. Other agencies with environmental justice components have similar problems.

General Environmental Policies with Civil Rights Relevance

Finally, the administration has adopted general environmental policies that affect minority and low-income communities seemingly without considering the civil rights consequences. For example, in 2002, President Bush signed the Small Business Liability Relief and Brownfields Revitalization Act of 2001 into law. Although on its face, the act seems like a viable plan to clean up abandoned and contaminated sites and redevelop them for commercial and residential purposes, it fails to account for the reality that white neighborhoods, regardless of economic status, generally fare better under such programs than minority communities. White communities experience faster cleanup, better results, and harsher assessments against those failing to clean them up as expected, despite that many abandoned industrial sites exist in minority and low-income communities. Moreover, the act removes legal responsibility for contaminated sites from prospective buyers, contiguous property owners, and "innocent" landowners. In so doing, it exempts small businesses from fines when they or their agents, for example, dispose of trash at waste sites and provides incentive for redevelopment. However, this exemption raises concern that removing responsibilities from companies will also remove legal remedies available to those most hurt by contamination, which are disproportionately minority and low-income communities. Furthermore, redevelopment in minority communities is not always successful because there is no vigorous enforcement of regulations that establish redevelopment standards.

Similarly, the Bush administration has supported the proposed Clean Skies Act despite that the proposal does not provide environmental justice safeguards. The Clean Skies Act would repeal the Clean Air Act's requirement that facilities install updated pollution control equipment when expanding their capacities. The Clean Skies Act also expands emissions allowances within geographic regions to include nitrogen oxides and mercury, in addition to sulfur dioxide. Under the Clean Skies Act, power plants must have one allowance for every ton of pollution they emit. A power plant that reduces emissions can sell unused allowances on the open market to another power plant. Consequently, the proposed act permits power plants to emit unlimited amounts of nitrogen oxides and mercury provided they purchase allowances from environmentally sound plants.

According to researchers, allowing power plants to increase their emissions may heighten the concentration of toxins in predominantly minority and low-income communities because many of the plants that purchase allowances are located in such areas. For example, in San Francisco, 87 percent of the pollution allowances produced by environmentally sound power plants were purchased by refineries and power plants in heavily industrialized and predominately minority and low-income communities.

In sum, the Bush administration has not adequately addressed environmental justice issues. Despite developing environmental justice goals, EPA has not created methods to measure the administration's progress toward them. Although EPA has reduced its backlog of Title VI complaints, concerns have been raised about those that were dismissed or rejected. The administration also does not recognize or acknowledge that minority and low-income communities are disproportionately exposed to environmental pollutants. Despite both the administration's position and whether municipalities and states are intentionally carrying out discriminatory policies, numerous studies reveal that minority communities are home to a disproportionate share of health hazards and waste sites. Minority and low-income people are victimized, and the Bush administration must acknowledge that environmental injustice exists.

Source: U.S. Commission on Civil Rights. Office of Civil Rights Evaluation. *Redefining Rights in America: The Civil Rights Record of the George W. Bush Administration, 2001–2004.* Draft Report. Washington, DC: U.S. Commission on Civil Rights, September 2004, pp. 72–79.

Environmental Justice: EPA Should Devote More Attention to Environmental Justice When Developing Clean Air Rules (2005)

In 2005, at the request of Hilda L. Solis (D-CA), ranking member on the House Subcommittee on Environment and Hazardous Materials of the Committee on Energy and Commerce, the Government Accountability Office (GAO) conducted a study to see how effectively the EPA incorporated principles of environmental justice in its rulemaking about clean air regulations. The GAO's conclusions and recommendations are listed below.

Conclusions

We found some evidence that EPA officials considered environmental justice when drafting or finalizing the three clean air rules we examined. During the drafting of the three rules, even when the workgroups discussed environmental justice, their capability to identify potential concerns may have been limited by a lack of guidance, training, and involvement of EPA's environmental justice coordinators. It is important that EPA thoroughly consider environmental justice because the states and other entities, which generally have the primary permitting authority, are not subject to Executive Order 12898.

EPA's capability to identify environmental justice concerns through economic reviews also appears to be limited. More than 10 years have elapsed since the executive order directed federal agencies, to the extent practicable and permitted by law, to identify and address the disproportionately high and adverse human health or environmental effects of their programs, policies, and activities. However, EPA apparently does not have sufficient data and modeling techniques to be able to distinguish localized adverse impacts for a specific community. For example, EPA has not agreed upon the complete list of data that would be needed to perform an environmental justice analysis. This suggests that, although EPA has developed general guidance for considering environmental justice, it has not established specific modeling techniques for assessing the potential environmental justice implications of any clean air rules. In addition, by not including a discussion of environmental justice in all of the economic reviews, EPA decision makers may not have been fully informed about the environmental justice impacts of all the rules.

Finally, even though members of the public commented about two rules' potential to increase refinery emissions—potential environmental justice issues, (1) in one case, EPA did not provide a response and (2) in the other case, it did not explain the basis for its response, such as the rationale for its beliefs and the data on which it based its beliefs. While these may not have been significant comments requiring a response, EPA's public involvement policy calls for EPA to provide responses when feasible, and this policy does not appear to distinguish comments on Advanced Notices of Proposed Rulemaking from comments on proposed rules.

Recommendations for Executive Action

In order to ensure that environmental justice issues are adequately identified and considered when clean air rules are being drafted and finalized, we recommend that the EPA Administrator take the following four actions:

- ensure that the workgroups devote attention to environmental justice while drafting and finalizing clean air rules;
- enhance the workgroups' ability to identify potential environmental justice issues through such steps as (1) providing workgroup members with guidance and training to help them identify potential environmental justice problems and (2) involving environmental justice coordinators in the workgroups when appropriate;
- improve assessments of potential environmental justice impacts in economic reviews by identifying the data and developing the modeling techniques that are needed to assess such impacts; and
- direct cognizant officials to respond fully, when feasible, to public comments on environmental justice, for example, by better explaining the rationale for EPA's beliefs and by providing its supporting data.

Source: U.S. Government Accountability Office. *Environmental Justice: EPA Should Devote More Attention to Environmental Justice When Developing Clean Air Rules.* Report GAO-05-289. Washington, DC: Government Accountability Office, July 2005, 23–24.

Toxic Wastes and Race at Twenty (2007)

In 2007, the Justice and Witness Ministries of the United Church of Christ published a report on the status of the environmental justice movement in the United States, an updated version of its groundbreaking 1987 report. In the executive summary of the report, its authors presented the key findings and conclusions of their study.

Key Findings

The application of these new methods, which better determine where people live in relation to where hazardous sites are located, reveals that racial disparities in the distribution of hazardous wastes are greater than previously reported. In fact, these methods show that people of color make up the majority of those living in host neighborhoods within 3 kilometers (1.8 miles) of the nation's hazardous waste facilities. Racial and ethnic disparities are prevalent throughout the country.

National Disparities.   More than nine million people (9,222,000) are estimated to live in circular host neighborhoods within 3 kilometers of the nation's 413 commercial hazardous waste facilities. More than 5.1 million people of color, including 2.5 million Hispanics or Latinos, 1.8 million African Americans, 616,000 Asians/Pacific Islanders and 62,000 Native Americans live in neighborhoods with one or more commercial hazardous waste facilities. Host neighborhoods of commercial hazardous waste facilities are 56% people of color whereas non-host areas are 30% people of color. Percentages of African Americans, Hispanics/Latinos and Asians/Pacific Islanders in host neighborhoods are 1.7, 2.3 and 1.8 times greater (20% vs. 12%, 27% vs. 12%, and 6.7% vs. 3.6%), respectively. Poverty rates in the host neighborhoods are 1.5 times greater than non-host areas (18% vs. 12%).

Neighborhoods with Clustered Facilities.   Neighborhoods with facilities clustered close together have higher percentages of people of color than those with non-clustered facilities (69% vs. 51%). Likewise, neighborhoods with clustered facilities have disproportionately high poverty rates. Because people of color and the poor are highly concentrated in neighborhoods with multiple facilities, they continue to be particularly vulnerable to the various negative impacts of hazardous waste facilities.

EPA Regional Disparities.   Racial disparities for people of color as a whole exist in nine out of 10 U.S. EPA regions (all except Region 3). Disparities in people of color percentages between host neighborhoods and non-host areas are greatest in: Region 1, the Northeast (36% vs. 15%); Region 4, the southeast (54% vs. 30%); Region 5, the Midwest (53% vs. 19%); Region 6, the South (63% vs. 42%); and Region 9, the southwest (80% vs. 49%). For Hispanics, African Americans and Asians/Pacific Islanders, statistically significant disparities exist in the majority or vast majority of EPA regions. The pattern of people of color being especially concentrated in areas where facilities are clustered is also geographically widespread throughout the country.

State Disparities.   Forty of the 44 states (90%) with hazardous waste facilities have disproportionately high percentages of people of color in circular host neighborhoods within 3 kilometers of the facilities. States with the 10 largest differences in people of color percentages between host neighborhoods and non-host areas include (in descending order by the size of the differences): Michigan (66% vs. 19%), Nevada (79% vs. 33%), Kentucky (51% vs. 10%), Illinois (68% vs. 31%), Alabama (66% vs. 31%), Tennessee (54% vs. 20%), Washington (53% vs. 20%), Kansas (47% vs. 16%), Arkansas (52% vs. 21%) and California (81% vs. 51%). Thirty-five states have socioeconomic disparities, i.e., in poverty rates. In these states, the average poverty rate in host neighborhoods is 18% compared to 12% in non-host areas.

Metropolitan Disparities. In metropolitan areas, where four of every five hazardous waste facilities are located, people of color percentages in hazardous waste host neighborhoods are significantly greater than those in non-host areas (57% vs. 33%). Likewise, the nation's metropolitan areas show disparities in percentages of African Americans, Hispanics/Latinos and Asians/Pacific Islanders, 20% vs. 13%, 27% vs. 14% and 6.8% vs. 4.4%, respectively. Socioeconomic disparities exist between host neighborhoods and non-host areas, with poverty rates of 18% vs. 12%, respectively. One hundred and five of the 149 metropolitan areas with facilities (70%) have host neighborhoods with disproportionately high percentages of people of color, and 46 of these metro areas (31%) have majority people of color host neighborhoods.

Continuing Significance of Race. In 1987, Toxic Wastes and Race in the United States found race to be more important than socioeconomic status in predicting the location of the nation's commercial hazardous waste facilities. In 2007, our current study results show that race continues to be a significant and robust predictor of commercial hazardous waste facility locations when socioeconomic factors are taken into account.

Conclusions

Twenty years after the release of Toxic Wastes and Race, significant racial and socioeconomic disparities persist in the distribution of the nation's commercial hazardous waste facilities. Although the current assessment uses newer methods that better match where people and hazardous waste facilities are located, the conclusions are very much the same as they were in 1987.

Race matters. People of color and persons of low socioeconomic status are still disproportionately impacted and are particularly concentrated in neighborhoods and communities with the greatest number of facilities. Race continues to be an independent predictor of where hazardous wastes are located, and it is a stronger predictor than income, education and other socioeconomic indicators. People of color now comprise a majority in neighborhoods with commercial hazardous waste facilities, and much larger (more than two-thirds) majorities can be found in neighborhoods with clustered facilities. African Americans, Hispanics/Latinos and Asian Americans/Pacific Islanders alike are disproportionately burdened by hazardous wastes in the U.S.

Place matters. People of color are particularly concentrated in neighborhoods and communities with the greatest number of hazardous waste facilities, a finding that directly parallels that of the original UCC report. This current appraisal also reveals that racial disparities are widespread throughout the country, whether one examines EPA regions, states or metropolitan areas, where the lion's share of facilities is located. Significant racial and socioeconomic disparities exist today despite the considerable societal attention to the problem noted in this report. These findings raise serious questions about the ability of current policies and institutions to adequately protect people of color and the poor from toxic threats.

Unequal protection places communities of color at special risk. Not only are people of color differentially impacted by toxic wastes and contamination, they can expect different responses from the government when it comes to remediation—as clearly seen in the two case studies in Post-Katrina New Orleans and in Dickson County, Tennessee. Thus, it does not appear that existing environmental, health and civil rights laws and local land use controls have been adequately applied or adapted to reducing health risks or mitigating various adverse impacts to families living in or near toxic "hot spots."

Polluting industries still follow the path of least resistance. For many industries it is a "race to the bottom," where land, labor and lives are cheap. It's about profits and the "bottom line." Environmental "sacrifice zones" are seen as the price of doing business. Vulnerable communities, populations and individuals often fall between the regulatory cracks. They are in many ways "invisible" communities. The environmental justice movement served to make these disenfranchised communities visible and vocal.

The current environmental protection apparatus is "broken" and needs to be "fixed." The current environmental protection system fails to provide equal protection to people of color and low-income communities. Various levels of government have been slow to respond to environmental health threats from toxic waste in communities of color. The mission of the United States Environmental Protection Agency (EPA) was never to address environmental policies and practices that result in unfair, unjust and inequitable outcomes. The impetus for change came from grassroots mobilization that views environmental protection as a basic right, not a privilege reserved for a few who can "vote with their feet" and escape from or fend off locally undesirable land uses—such as landfills, incinerators, chemical plants, refineries and other polluting facilities.

Slow government response to environmental contamination and toxic threats unnecessarily endangers the health of the most vulnerable populations in our society. Government officials have knowingly allowed people of color families near Superfund sites, other contaminated waste sites and polluting industrial facilities to be poisoned with lead, arsenic, dioxin, TCE, DDT, PCBs and a host of other deadly chemicals. Having the facts and failing to respond is explicitly discriminatory and tantamount to an immoral "human experiment."

Clearly, the environmental justice movement over the last two decades has made a difference in the lives of people of color and low-income communities that are overburdened with environmental pollution. After years of intense study, targeted research, public hearings, grassroots organizing, networking and movement building, environmental justice struggles have taken center stage. However, community leaders who have been on the front line for justice for decades know that the lethargic, and too often antagonistic, government response to environmental emergencies in their communities is not the exception but the general rule. They have come to understand that waiting for the government to respond can be hazardous to their health and the health of their communities.

In fact, the U.S. EPA, the governmental agency millions of Americans look to for protection, has mounted an all-out attack on environmental justice and environmental justice principles established in the early 1990s. Moreover, the agency has failed to implement the Environmental Justice Executive Order 12898 signed by President Bill Clinton in 1994 or adequately apply Title VI of the Civil Rights Act.

Source: Justice and Witness Ministries (United Church of Christ). *Toxic Wastes and Race at Twenty 1987–2007,* x–xiii. Copyright March 2007 United Church of Christ. The Commission for Racial Justice (United Church of Christ). All rights reserved. Used by permission.

S. 2549: Environmental Justice Renewal Act (2008)

Environmental justice legislation has been introduced into the U.S. Senate and the U.S. House of Representatives a number of times in the past two decades. None of the bills has ever been passed by either body. The most recent form of the legislation was introduced by Sen. Hillary Rodham Clinton (D-NY) on January 23, 2008. Senator Clinton's bill was referred to the Committee on Environment and Public Works, but no action was taken on it during the 110th Congress. Portions of the bill are reprinted here.

A Bill

To require the Administrator of the Environmental Protection Agency to establish an Interagency Working Group on Environmental Justice to provide guidance to Federal agencies on the development of criteria for identifying disproportionately high and adverse human health or environmental effects on minority populations and low-income populations, and for other purposes.

Be it enacted by the Senate and House of Representatives of the United States of America in Congress assembled,

Section 1. Short Title

This Act may be cited as the "Environmental Justice Renewal Act."

Section 2. Definitions

[This section defines key terms used in the bill.]

Section. 3. Interagency Working Group on Environmental Justice

(a) Establishment—Not later than 30 days after the date of enactment of this Act, the Administrator shall establish a working group to be known as the "Interagency Working Group on Environmental Justice."

(b) Purposes—The purposes of the Working Group are—

(1) to provide guidance to Federal agencies on the development of the guidance document under subsection (f)(1) for identifying disproportionately high and adverse human health or environmental effects on—

(A) minority populations; and

(B) low-income populations;

(2) to coordinate with, provide guidance to, and serve as a clearinghouse for, each Federal agency during the development by each Federal agency of an environmental justice strategy;

(3) to ensure that the administration, interpretation, and enforcement of each applicable program, activity, and policy of each Federal agency is undertaken in a manner that minimizes or eliminates disproportionately high and adverse human health or environmental effects on racial minority, ethnic minority, or low-income populations;

(4) to assist in the coordination of research conducted by, and stimulate cooperation among—

(A) the Agency;

(B) the Department of Health and Human Services;

(C) the Department of Housing and Urban Development;

(D) the Department of Transportation; and

(E) any other Federal agency that conducts research or any other activity relating to the study of human health and environmental research and analysis;

(5) to assist in the coordination of data collection activities conducted by each Federal agency described in paragraph (4);

(6) to examine each study and available data with respect to issues relating to environmental justice in existence as of the date of enactment of this Act;

(7) to hold public meetings to conduct fact-finding, receive public comments, and conduct inquiries concerning issues relating to environmental justice, the summaries of the comments and recommendations from which shall be made available to the public;

(8) to develop interagency model projects on issues relating to environmental justice that evidence cooperation among Federal agencies;

(9) to engage in regular consultation with the Advisory Council, but not less than once per year;

(10) to assess and review the activities of the Federal Government (including any policy or program of the Federal Government in existence as of the date of enactment of this Act) to minimize and eliminate disproportionately high and adverse human health or environmental effects on racial minority, ethnic minority, or low-income populations; and

(11) to seek advice from community-based organizations and academic experts who are engaged in environmental justice research and other activities.

(c) Composition—The Working Group shall be composed of—

(1) the Administrator (or a designee);

(2) the Secretary of Defense (or a designee);

(3) the Secretary of Health and Human Services (or a designee);

(4) the Secretary of Housing and Urban Development (or a designee);

(5) the Secretary of Labor (or a designee);

(6) the Secretary of Agriculture (or a designee);

(7) the Secretary of Transportation (or a designee);

(8) the Attorney General (or a designee);

(9) the Secretary of the Interior (or a designee);

(10) the Secretary of Commerce (or a designee);

(11) the Secretary of Energy (or a designee);

(12) the Secretary of Homeland Security (or a designee);

(13) the Director of the Office of Management and Budget (or a designee);

(14) the Director of the Office of Science and Technology Policy (or a designee);

(15) the Deputy Assistant to the President for Environmental Policy (or a designee);

(16) the Assistant to the President for Domestic Policy (or a designee);

(17) the Director of the National Economic Council (or a designee);

(18) the Chairman of the Council of Economic Advisers (or a designee); and

(19) any other official of the Federal Government that the President may designate.

(d) Chairperson—The President (or a designee) shall serve as the Chairperson of the Working Group.

(e) Report to President—The Working Group shall report to the President through—

(1) the Deputy Assistant to the President for Environmental Policy; or

(2) the Assistant to the President for Domestic Policy.

(f) Uniform Consideration Guidance Document—

(1) IN GENERAL—To ensure that there is a common level of understanding of terminology used in dealing with environmental justice issues, not later than 1 year after the date of enactment of this Act, the Working Group shall develop and publish in the *Federal Register* a guidance document that outlines the ways in which the following considerations will be taken into account in defining communities as environmental justice communities:

(A) A disproportionate burden of adverse human health or environmental impacts, or the potential for those impacts.

(B) Aggregation of risk.

(C) Cumulative sources of risk.

(D) Additional elements in the community that indicate vulnerability, such as a high incidence of disease and lower income levels.

(E) The capacity of communities to address environmental concerns.

[The remainder of this section deals with "housekeeping" issues, such as public comment, reports, and documentation.]

Section 4. Responsibilities of Federal Agencies

(a) Conduct of Programs—Each Federal agency shall conduct each program, policy and activity of the Federal agency that adversely affects or has the potential to affect human health or the environment in a manner that ensures that each program, policy, and activity does not have an effect of excluding any individual from participating in, denying any individual the benefits of, or subjecting any individuals to discrimination or disparate impact under any program, policy, or activity of the Federal agency because of the race, color, national origin, or income level of the individual.

(b) Federal Agency Environmental Justice Strategies—

(1) DEVELOPMENT—Each Federal agency that participates in the Working Group shall develop an agency-wide environmental justice strategy that—

(A) identifies and addresses any disproportionately high or adverse human health or environmental effects of each program, policy, and activity of the Federal agency on—

(i) minority populations; and

(ii) low-income populations; and

(B) complies with each requirement described in paragraph (2).

(2) CONTENTS—Each environmental justice strategy developed by a Federal agency under paragraph (1) shall contain—

(A) an assessment that identifies each program, policy, planning and public participation process, enforcement activity, and rulemaking procedure relating to human health or the environment that the Federal agency determines should be revised—

(i) to ensure the enforcement by the Federal agency of each health or environmental statute relating to—

(I) minority populations; and

(II) low-income populations;

(ii) to ensure greater and more meaningful public participation;

(iii) to improve the conduct by the Federal agency of research and data collection activities relating to the health and environment of—

(I) minority populations; and

(II) low-income populations;

(iv) to facilitate the identification of differential development patterns and access to natural resources among—

(I) minority populations; and

(II) low-income populations;

(v) to integrate environmental justice into the activities of the Federal agency; and

(vi) to ensure that the Federal agency undertakes activities to reduce or eliminate disproportionately high and adverse human health or environmental effects on racial minority, ethnic minority, and low-income populations; and

(B) a timetable for the completion of—

(i) each revision identified under subparagraph (A); and

(ii) an assessment of the economic and social implications of each revision identified under subparagraph (A).

[The rest of this section deals with details as to the adoption and implementation of agency plans for environmental justice actions.]

Section 5. Ombudsmen

[This section establishes a position of environmental justice ombudsman within the Environmental Protection Agency (EPA).]

Section 6. Training of Employees of Federal Agencies

[This section requires that employees of the EPA be offered an opportunity to receive training in environmental justice theory and practices.]

Section 7. Grant Programs

[This section authorizes the establishment of small grants, collaborative grants, interagency cooperative agreement programs, state grants, and community-based participatory research grant programs to carry out the provisions of the act and allocates up to $5 million for the implementation of most programs.]

Section 8. Environmental Justice Basic Training Program

(a) Establishment—The Administrator shall establish a basic training program to increase the capacity of residents of environmental justice communities to identify and address disproportionately high and adverse human health or environmental effects by providing culturally and linguistically appropriate—

(1) training relating to—

(A) basic and advanced techniques for the detection, assessment, and evaluation of the effects of hazardous substances on human health;

(B) methods to assess the risks to human health presented by hazardous substances;

(C) methods and technologies to detect hazardous substances in the environment; and

(D) basic biological, chemical, and physical methods to reduce the quantity and toxicity of hazardous substances; and

(2) short courses and continuation education programs for residents of communities who are located in close proximity to hazardous substances to provide—

(A) education relating to—

(i) the proper manner to handle hazardous substances;

(ii) the management of facilities at which hazardous substances are located (including facility compliance protocols); and

(iii) the evaluation of the hazards that facilities described in clause (ii) pose to human health; and

(B) training on environmental and occupational health and safety with respect to the public health and engineering aspects of hazardous waste control.

(b) Grant Program—

[This section describes the grant program established to support training in environmental justice.]

Section 9. National Environmental Justice Advisory Council

[This section provides for a national environmental justice advisory council consisting of representatives from community-based organizations, state and local governments, Indian tribes and other indigenous groups, nongovernmental and environmental organizations, and private-sector organizations, as well as experts from a number of fields related to environmental justice.]

Section 10. Environmental Justice Clearinghouse

[This section establishes a reference center consisting of "culturally and linguistically appropriate materials" that includes:]

(1) information describing the activities conducted by the Environmental Protection Agency to address issues relating to environmental justice;

(2) copies of training materials provided by the Administrator to help individuals and employees understand and carry out environmental justice activities;

(3) links to webpages that describe environmental justice activities of other Federal agencies;

(4) a directory of individuals who possess technical expertise in issues relating to environmental justice;

(5) a directory of nonprofit and community-based organizations that address issues relating to environmental justice at the local, State, and Federal levels (with particular emphasis given to nonprofit and community-based organizations that possess the capability to provide advice or technical assistance to environmental justice communities); and

(6) any other appropriate information, as determined by the Secretary.

[The remainder of this section deals with "housekeeping" issues about the clearinghouse.]

[The final three sections of the bill deal with public meetings, supplemental environmental projects for environmental justice communities, and evaluation by the comptroller general of the United States.]

Source: S. 2549. Environmental Justice Renewal Act (introduced in Senate). http://thomas.loc.gov/cgi-bin/query/z?c110:S.2549:.

Data

Some of the most complete data on environmental inequities come from the 2007 report by the Justice and Witness Ministries of the United Church of Christ, Toxic Wastes and Race at Twenty 1987–2007. *The following tables are reprinted from that report.*

TABLE 6.2
Racial and Socioeconomic Characteristics of
People Living Near Hazardous Waste Facilities

	Within 1 km	Between 1 km and 3 km	Between 3 km and 5 km	Beyond 5 km
Population				
Total population (in 1,000s)	845	7,828	14,101	225,936
Population density (persons per square kilometer)	690	840	810	24
Race/ethnicity				
People of color (%)	47.7	46.1	35.7	22.2
African American (%)	20.6	20.4	20.6	11.2
Hispanic (%)	23.1	20.4	18.1	7.8
Asian/Pacific Islander (%)	4.4	5.4	5.3	2.7
Native American (%)	0.6	0.6	0.5	0.8
Socioeconomic characteristics				
Poverty rate (%)	20.1	18.3	16.9	12.7
Mean household income	$31,192	$33,318	$36,920	$38,745
Mean housing value	$93,985	$102,594	$111,915	$111,956

Source: Justice and Witness Ministries (United Church of Christ). *Toxic Wastes and Race at Twenty 1987–2007*, p. 13. Copyright March 2007 United Church of Christ. The Commission for Racial Justice (United Church of Christ). All rights reserved. Used by permission.

TABLE 6.3

Racial and Socioeconomic Disparities between Host Neighborhoods and Non-Host Areas for the Nation's 413 Commercial Hazardous Waste Facilities (1990 and 2000 Census)

	2000				1990			
	Host	Non-Host	Difference	Ratio	Host	Non-Host	Difference	Ratio
Population								
Total population (in 1,000s)	9,222	272,200	262,979	0.03	8,673	240,037	231,364	0.04
Population density	870	29.7	840	29.0	820	25.1	790	27.3
Race/ethnicity								
People of color	55.9%	30.0%	25.9%	1.86	46.2%	23.4%	22.8%	1.97
African American	20.0%	11.9%	8.0%	1.67	20.4%	11.7%	8.7%	1.74
Hispanic or Latino	27.0%	12.0%	15.0%	2.25	20.7%	8.4%	12.3%	2.47
Asian/Pacific Islander	6.7%	3.6%	3.0%	1.83	5.3%	2.8%	2.5%	1.88
Native American	0.7%	0.9%	0.2%	0.77	0.6%	0.8%	0.3%	0.68
Socioeconomics								
Poverty rate	18.3%	12.2%	6.1%	1.50	18.5%	12.9%	5.6%	1.43
Mean household income	$48,234	$56,912	$8,678	0.85	$33,115	$38,639	$5,524	0.86
Mean owner-occupied housing value	$135,510	$159,536	$24,025	0.85	$101,774	$111,954	$10,180	0.91
With four-year college degree	18.5%	24.6%	6.1%	0.75	15.4%	20.5%	5.1%	0.7
Professional white collar occupations	28.0%	33.8%	5.8%	0.83	21.8%	26.6%	4.8%	0.82
Employed in blue-collar occupations	27.7%	24.0%	3.7%	1.15	30.0%	26.1%	3.9%	1.15

TABLE 6.4
People of Color Percentages for
Host Neighborhoods and Non-Host Areas by EPA Region

	Number of Facilities	Host Neighborhoods	Non-Host Areas	Difference	Ratio
U.S. total	413	55.9%	30.0%	25.9%	1.86
Region 1	23	36.3%	15.0%	21.3%	2.43
Region 2	32	51.5%	36.0%	15.6%	1.43
Region 3	35	23.2%	24.5%	1.33%	0.95
Region 4	67	54.3%	30.4%	23.8%	1.78
Region 5	85	52.6%	18.8%	33.8%	2.80
Region 6	61	62.7%	41.8%	20.9%	1.50
Region 7	32	29.1%	13.4%	15.7%	2.17
Region 8	15	31.2%	18.2%	13.0%	1.72
Region 9	55	80.5%	49.4%	31.1%	1.63
Region 10	8	38.9%	19.1%	19.9%	2.04

Source: Justice and Witness Ministries (United Church of Christ). *Toxic Wastes and Race at Twenty 1987–2007*, p. 13. Copyright March 2007 United Church of Christ. The Commission for Racial Justice (United Church of Christ). All rights reserved. Used by permission.

TABLE 6.5

Socioeconomic Disparities between Host Neighborhoods and Non-Host Areas by EPA Region

EPA region	Poverty Rates				Mean Household Income				Mean Housing Value			
	Host	Non-host	Diff.	Ratio	Host	Non-host	Diff.	Ratio	Host	Non-host	Diff.	Ratio
Region 1	15.7%	8.7%	7.0%	1.80	$48,368	$65,296	$16,928	0.74	$143,840	$202,102	$58,261	0.71
Region 2	19.4%	12.3%	7.1%	1.57	$50,793	$66,137	$15,344	0.77	$171,083	$202,579	$31,496	0.84
Region 3	12.6%	10.7%	1.9%	1.18	$47,493	$57,479	$9,986	0.83	$97,971	$139,278	$41,307	0.70
Region 4	15.7%	13.7%	2.0%	1.15	$45,811	$50,931	$5,120	0.90	$97,673	$118,962	$21,288	0.82
Region 5	19.4%	9.6%	9.7%	2.01	$44,933	$56,955	$12,022	0.79	$103,812	$137,470	$33,658	0.76
Region 6	18.8%	16.0%	2.8%	1.18	$45,072	$50,616	$5,545	0.89	$83,602	$101,518	$17,916	0.82
Region 7	15.0%	10.4%	4.7%	1.45	$44,084	$50,308	$6,224	0.88	$84,028	$106,808	$22,780	0.79
Region 8	14.8%	10.3%	4.4%	1.43	$40,801	$55,413	$14,612	0.74	$105,286	$163,390	$58,104	0.64
Region 9	20.7%	13.5%	7.2%	1.54	$52,947	$64,146	$11,199	0.83	$218,576	$246,673	$28,096	0.89
Region 10	10.9%	11.0%	0.1%	0.99	$55,599	$55,889	$290	0.99	$180,716	$179,522	$1,193	1.01

Source: Justice and Witness Ministries (United Church of Christ). *Toxic Wastes and Race at Twenty 1987–2007*, p. 13. Copyright March 2007 United Church of Christ. The Commission for Racial Justice (United Church of Christ). All rights reserved. Used by permission.

7

Directory of Organizations

The number of organizations interested in the issue of environmental justice continues to grow with each passing year. Some of these organizations have broad-based interests, ranging from social, economic, and political issues faced by nonwhite and/or lower-income populations to environmental issues of every kind. Other organizations are focused more specifically on environmental racism, environmental equity, and environmental justice on a local, regional, or national level. Probably the single most useful resource listing these organizations is *People of Color Environmental Groups Directory*, compiled and edited by Robert D. Bullard. The current edition was published in 2000 and is an update of the original 1992 version of the directory. It contains information on more than 400 groups in 45 states, the District of Columbia, and Puerto Rico. That number represents a doubling of the list of groups listed in the 1992 directory. The book includes an excellent general introduction to environmental justice, a listing of groups, an annotated bibliography, and additional information on environmental justice.

Given the scope of the environmental justice movement, the present chapter can do no more than list some of the major national organizations with a significant interest in environmental justice, along with a sampling of regional and local groups of special interest. The chapter is divided into two major sections: (1) governmental agencies and (2) private and nongovernmental organizations.

Governmental Organizations

Federal Highway Administration (FHA)
Web site: http://www.fhwa.dot.gov/environment/ej2.htm
e-mail: brenda.kragh@dot.gov

Federal Transit Administration (FTA)
Web site: http://www.fta.dot.gov/
e-mail: FTA.ADAAssistance@dot.gov

Under the mandate of Executive Order 12898, the Federal Highway Administration and the Federal Transit Administration now include a number of environmental justice considerations in their planning, grant making, and program development. These considerations include the development of transit systems that will be of greatest benefit to all people, the construction of such systems with the least possible impact on citizens of all color and economic classes, the integration of transit systems into community environments, an increase in the involvement of individual citizens and citizen groups in transit planning, and improvement of data collection and analysis on the impact of transit systems on communities.

National Environmental Justice Advisory Council. *See* **U.S. Environmental Protection Agency, Office of Environmental Justice**

National Institute of Environmental Health Sciences (NIEHS)
Web site: http://www.niehs.nih.gov/research/supported/
    programs/justice/
e-mail: ofallon@niehs.nih.gov

In 1994, the National Institute of Environmental Health Sciences (NIEHS) created the Environmental Justice: Partnerships for Communication Program. The program provides grants to local consortia that consist of three components: a community organization, an environmental health researcher, and a health care professional. Members of the consortia work together to develop methods by which local citizens can be involved in research on environmental issues relating to their own lives.

Office of Environmental Justice (EPA). *See* **U.S. Environmental Protection Agency, Office of Environmental Justice**

Office of Solid Waste and Emergency Response. *See* **U.S. Environmental Protection Agency, Office of Solid Waste and Emergency Response**

U.S. Department of the Interior (USDI) Assistant Secretary for Policy, Management, and Budget
Office of Environmental Policy and Compliance (OEPC)
Web site: http://www.doi.gov/oepc/ej_goal1.html
e-mail: Emily_Joseph@ios.doi.gov

In 1994, the Department of the Interior (USDI) adopted four goals to guide its implementation of Executive Order 12898: (1) to involve low-income and minority communities in making environmental decisions and to provide those communities with information about the decisions; (2) to provide members of low-income and minority communities with the type of training that will help them reduce their exposure to hazardous materials; (3) to expand research and data gathering on problems of environmental inequities; and (4) to expand departmental connections with grassroots organizations, environmental groups, businesses, academic institutions, and other organizations to work toward environmental justice in fields for which the USDI is responsible. The department's Web site also provides a detailed discussion of the ways in which the goals have been met by the its major divisions, the Bureau of Indian Affairs, Bureau of Land Management, Bureau of Mines, Bureau of Reclamation, Minerals Management Service, National Biological Survey, National Park Service, Office of Surface Mining, U.S. Fish and Wildlife Service, and U.S. Geological Survey. As even a cursory review of these titles suggest, the Department of the Interior is intimately involved in many of the controversies characteristic of the environmental justice movement.

U.S. Environmental Protection Agency (EPA)
Office of Environmental Justice (OEJ)
Web site: http://www.epa.gov/environmentaljustice/
e-mail: See "Contact Us" page on Web site

The Environmental Protection Agency (EPA) created the Office of Environmental Justice (OEJ) in 1992 to ensure that concerns about environmental equality are integrated with all parts of the agency's mission and activities. It consists of two major divisions, the Executive Steering Committee, representing each headquarter's office and region, and the Environmental Justice Coordinators Council, responsible for the day-to-day operations of the office. In 1993, the EPA also formed the National Environmental Justice Advisory Council consisting of about 26 members who represent academia, community-based organizations, industry and business, nongovernmental and environmental groups, state and local governments, and tribal governments and indigenous organizations. The council is responsible for analyzing environmental justice issues and suggesting new and innovative solutions for resolving those problems.

U.S. Environmental Protection Agency (EPA)
Office of Solid Waste and Emergency Response (OSWER)
1200 Pennsylvania Avenue NW
Washington, DC 20460
MC-5101T
Telephone: (202) 529-7534
Web site: http://www.epa.gov/oswer/ej/index.html

The EPA's Office of Solid Waste and Emergency Response (OSWER) has regulatory responsibility for some of the primary areas in which advocates for environmental justice are concerned, areas such as brownfields cleanup and redevelopment; Superfund programs; underground storage tanks; emergency prevention, preparedness, and response; and federal facilities restoration. In response to Executive Order 12898, OSWER has developed a series of ongoing programs for ensuring that waste disposal projects do not produce a disproportionate risk for minority and low-income populations. The agency's Web site lists a number of examples of its programs that have worked to achieve this goal over the last decade or more.

Nongovernmental Organizations

National and International

African American Environmentalist Association (AAEA)
Web site: http://www.aaenvironment.com
e-mail: africanamericanenvironmentalist@msn.com

The African American Environmentalist Association (AAEA) was founded in 1985 by Norris McDonald to protect the environment, promote the efficient use of natural resources, increase the participation of African Americans in the environmental movement, and aid African American communities in dealing with environmental problems impacting their own communities. The organization has a wide range of interests, including toxic chemicals, nuclear power plants, stem cell research, nanotechnology, global climate change, and alternative energy sources. AAEA is an affiliate of a larger association also founded by McDonald, the Center for Environment, Commerce & Energy.

Alliance for Nuclear Accountability (ANA)
Web site: http://www.ananuclear.org/
e-mail: See "Contact Us" page on Web site

The alliance was founded in 1987 as the Military Production Network. It is currently a network of 35 local, regional, and national organizations representing communities located near nuclear weapons sites and radioactive waste dumps. The major issues with which ANA deals are the health issues of nuclear weapons construction, testing, and use; disposal and storage of nuclear wastes; and budget issues related to nuclear weapons manufacture and nuclear waste disposal. The organization maintains an extensive online library of articles dealing with nuclear weapons and nuclear wastes.

Americans for Indian Opportunity (AIO)
Web site: http://www.aio.org/
e-mail: mlucero@aio.org

Americans for Indian Opportunity (AIO) was founded in 1970 by LaDonna Harris to serve as a catalyst for finding new concepts and new opportunities for Indian people. The organization works with tribal governments and peoples to find ways to deal with a changing world based on traditional tribal values. AIO's best-known program may be its American Indian Ambassador Program, a leadership program in which mid-level professionals are trained to improve the well-being and growth of their own communities. Other important AIO programs are its Indigenous Leaders Interactive System (ILIS™), which brings together tribal members to work on common problems; Intergovernmental Relations, which promotes ongoing projects involving federal and tribal leaders; and Community Service and Sharing Common Causes, through which the organization sponsors conferences, forums, and other meetings for tribal leaders, government officials, and representatives of other organizations.

Asian Pacific Environmental Network (APEN)
Web site: http://www.apen4ej.org/
e-mail: apen@apen4ej.org

The Asian Pacific Environmental Network (APEN) was organized in 1993 to provide a vehicle by which Asian Americans and Pacific Islanders could learn more about and begin to deal with environmental issues of particular concern to their neighborhoods. The organization works on three levels: direct organizing in local communities, building a network of Asian American and Pacific Islander organizations, and working in multiracial alliances to bring about national and regional change. An example of the organization's direct organizing focus is its Laotian Organizing Project (LOP). Created in 1995, LOP aims to assist a large Laotian community in California's Contra Costa County in combating the problems it faces as a result of its location in one of the most toxic areas anywhere in the United States.

Basel Action Network (BAN)
c/o Earth Economics
122 S. Jackson, Suite 320
Seattle, WA 98104
Telephone: (206) 652-5555
Web site: http://www.ban.org/

The Basel Action Network (BAN) is a program of Earth Economics, a 501(c)(3) charitable organization whose goal is to use economics to promote healthy ecosystems, communities, and economies. BAN is concerned with the economic and environmental inequities associated with the worldwide trade of hazardous materials, as well as the potential health and environmental hazards posed by such trade. The organization focuses its efforts on four major campaigns: the E-Waste Stewardship Project, a program aimed at reducing and, eventually, eliminating the export of electronic wastes from the United States to developing countries; Green Shipbreaking, an effort to ensure that U.S. shipping vessels do not carry hazardous materials; Zero Mercury Campaign, a program to develop an international treaty controlling the international transport of mercury; and Basel Ban Ratification, an effort to have the Basel Convention on the Control of Transboundary Movements of Hazardous Wastes and Their Disposal ratified by the United States.

Beyond Pesticides (BP)
Web site: http://www.beyondpesticides.org/
e-mail: info@beyondpesticides.org

Beyond Pesticides was founded in 1981 as the National Coalition against the Use of Pesticides to protect the general public against the health risks of many pesticides in common use by the agricultural industry. It has historically taken a two-pronged approach to meeting this objective: educating the public on the health risks posed by chemical pesticides and promoting the use of natural and other safe pesticides. The organization publishes two newsletters, *Pesticides and You,* with articles of general interest and published quarterly, and *Technical Report,* reviewing current information on pesticide issues and published monthly.

Center for Environmental Health (CEH)
Web site: http://www.cehca.org/
e-mail: See "Contact Us" page on Web site

The purpose of the Center for Environmental Health (CEH) is to eliminate the health risks posed by industrial chemicals to individuals, families, and communities. As part of its efforts, CEH has acknowledged the Principles of Environmental Justice and incorporates those principles into its own efforts to protect communities from the risks posed by chemical factories and waste sites. Through its Justice Fund, it provides grants to community-based, grassroots organizations striving to deal with environmental inequities in their areas. CEH has also taken legal action against industries and companies whose activities disproportionately affect low-income and ethnic communities. Its Web site lists a number of successes achieved over the past decade in combating environmental racism in California and other areas.

Center for Health, Environment and Justice (CHEJ)
Web site: http://www.chej.org
e-mail: chej@chej.org

One of the earliest events in the environmental justice movement involved an action by neighbors in the Love Canal district of New York State protesting the toxic environment on which their community was built. The leader of that movement, Lois Gibbs, later formed the Citizens Clearinghouse for Hazardous Waste in 1981 to provide support and advice for other communities faced with similar environmental problems. The organization later changed its name to the Center for Health, Environment and Justice, although it continues to provide the same services it did under its original name. It currently receives about 1,500 requests each year for assistance in protecting individuals and communities against toxic waste hazards, promoting children's health, and protecting consumers against hazardous products.

Center for Policy Alternatives (CPA)
Web site: http://www.cfpa.org/
e-mail: info@cfpa.org

The Center for Policy Alternatives (CPA) was established in 1975 as a nonprofit, nonpartisan organization promoting progressive policies in all 50 states. The center provides a network for state officials and activities to advance "proactive, progressive policy solutions" to common problems. The organization works at three levels: (1) providing leadership training programs for legislators; (2) developing policy tools on a wide variety of issues; and (3) building a network among like-minded legislators in all parts of the nation. CPA's model legislation for environmental justice has been a particularly helpful tool for legislators interested in environmental issues. Similar model

legislation is provided for issues such as campaign finance reform, child-care financing, family and work issues, health care,
and women and the economy.

Chemical Weapons Working Group (CWWG)
Web site: http://cwwg.org/
e-mail: kefcwwg@cwwg.org

The Chemical Weapons Working Group (CWWG) is a project of
the Kentucky Environmental Foundation (KEF) and consists of a
number of organizations working to oppose the incineration of
chemical weapons. These organizations are located throughout
the United States, Russia, and the Pacific Rim. Some members of
the network are Families Concerned about Nerve Gas Incineration (Alabama), Citizens for Safe Weapons Disposal (Colorado),
the Oregon Clearinghouse for Pollution Reduction, the Pacific
Asia Council of Indigenous Peoples, and the Center for Assistance to Ecological Initiatives (Russia). The organization was
created at a 1991 meeting sponsored by the KEF at which representatives of eight sites in the United States, the Pacific area, and
Russia agreed on certain general principles that would form the
basis of the new Chemical Weapons Working Group. Among
those principles was opposition to incinerating chemical
weapons and to moving them from place to place, and support
for the development of alternative methods of destroying these
weapons.

Citizens Coal Council (CCC)
Web site: http://www.citizenscoalcouncil.org/
e-mail: ccc@citizenscoalcouncil.org

The Citizens Coal Council (CCC) was organized in 1989 as a
grassroots federation of citizen groups working for social and environmental justice in the area of coal mining. The council consists of six member groups: Black Warrior Riverkeeper, Dakota
Resource Council, Friends of Hurricane Creek, West Virginia
Highlands Conservancy, Mountain Watershed Association, and
Save Our Cumberland Mountains. The organization has three
objectives: (1) to protect people, their homes and communities,
and the environment from damage caused by coal mining; (2) to
ensure enforcement of the federal Surface Mining Control and
Reclamation Act; and (3) to "help each other win our issues."

Community Toolbox for Children's Environmental Health
Web site: http://www.communitytoolbox.org/
e-mail: information@communitytoolbox.org

Community Toolbox for Children's Environmental Health works to eliminate environmental health risks (such as lead poisoning) faced by children, especially in high-risk communities. The organization currently focuses its efforts into four major programs. Organizational Development provides groups with the organizational tools they need to operate efficiently and effectively, including information on the establishment of an administrative structure, guidance on finance planning, methods of fund-raising, and effective use of technology. The Peer Network program encourages the development of connections among groups and individuals engaged in many groups and organizations. Leadership Development and Base Building teaches community leaders about principles of leadership skills, mentoring, constituency development, and community organizing. Policy, Media, and Research provides instruction on policy development and the use of media and research in community action.

CorpWatch
Web site: http://www.corpwatch.org/
e-mail: See "Contact" page on Web site

CorpWatch was founded in Oakland, California, in 1996 as TRAC—Transnational Resource and Action Center. It changed its name to CorpWatch in 2001 to make its overall mission clearer. The organization's mission is to investigate and expose corporate violations of human rights, environmental crimes, and fraud and corruption around the world. One of CorpWatch's many campaigns involved its participation in recent efforts to ensure the consideration of environmental justice issues in international programs to reduce and eliminate the shipment of hazardous wastes across national borders.

Diné CARE
Web site: http://dinecare.org/
e-mail: kiyaani@frontier.net

Diné CARE was formed in 1988 to defend the community of Dilkon in the southwestern region of the Navajo Nation, in the state of Arizona. After the tribal government had "signed off" on the construction of a toxic waste incinerator and dump near the town, some tribal members decided to take action against that decision. They educated themselves as to how they could take action and were eventually successful in preventing the incinerator and dump from being built. In 1990, Diné CARE cofounded the Indigenous Environmental Network. In the Navajo language, Diné means "the people."

Earthjustice
Web site: http://www.earthjustice.org/
e-mail: info@earthjustice.org

Earthjustice was created in 1971 as the Sierra Club Legal Defense Club. It changed its name in 1997 to better reflect the wide variety of programs in which it had become involved. Today, the organization claims to be "the leading nonprofit environmental law firm in the country." It is regularly involved in a number of different legal actions, ranging from efforts to improve the nation's smog standards, conducting ocean seismic surveys, fighting exemptions by the EPA for a number of industries that heavily pollute the environment, pressuring the EPA to develop regulations for mine cleanups by mining companies, and pressuring New York State to implement legislation for the cleanup of state brownfields.

Energy Justice Network (EJN)
Web site: http://www.energyjustice.net/
e-mail: See "Contact Us" page on the Web site

Energy Justice Network (EJN) was founded by Mike Ewall in 1999 after he heard a presentation by an environmental justice network in Louisiana. The purpose of the organization is to work for a clean energy, zero-emission, zero-waste future for all Americans. A major thrust of EJN's work is the total elimination of nuclear power generation in the country. The network takes as its guide the Principles of Environmental Justice adopted at the First National People of Color Environmental Leadership Summit held in 1991.

Environmental Justice and Climate Change Initiative (EJCCI)
Web site: http://www.ejcc.org/
e-mail: info@ejcc.org

Environmental Justice and Climate Change Initiative (EJCCI) is a coalition of organizations and networks concerned with issues of

environmental justice, climate justice, religion, policy development, and political advocacy. The organization's mission is to educate and energize the people of North America in the
development and implementation of just climate policies at both
domestic and international levels. In 2001, the organization developed "10 Principles for Just Climate Change Policies in the
U.S.," to be used as the foundation for ensuring that environmental justice is involved in the development and implementation of climate change policies in the United States. In December
2007, four members of EJCCI attended the UN Climate Change
Conference in Bali, Indonesia, to represent concerns for environmental justice in climate change discussions.

Environmental Justice and Health Union (EJHU)
Web site: http://www.ejhu.org/
e-mail: ejhu@ejhu.org

The Environmental Justice and Health Union (EJHU) is an independent project of the Center for Environmental Health. It was
formed in 2001 to improve the ability of environmental justice
and environmental health groups to work with each other. The
organization provides information about funding, partnerships, research, policy, and other tools through its monthly
newsletter *Catalyst* and a regularly updated Web site. Topics
currently discussed on the organization's Web site include disparities in the distribution of lead, PCBs, dioxins, and other environmental hazards; information on law, policy, training, and
other aspects of environmental justice and environmental
health issues; and background information on affirmative action and cultural competence.

Environmental Justice Coalition for Water (EJCW)
Web site: http://www.ejcw.org/
e-mail: debbie@ejcw.org

The Environmental Justice Coalition for Water (EJCW) was
formed in 1999 as a project of the Pacific Institute to deal with environmental problems caused by construction in the San Francisco Bay–Delta region of California. It is now a 501(c)(3)
nonprofit organization with a greatly expanded mission. It has
become a network of more than 50 grassroots organizations,

working to help local communities develop water-use programs equitable to all citizens. Among its constituent members are the Alliance for a Clean Waterfront; Butte Environmental Council; Californians for Pesticide Reform; Center for Race, Poverty, and the Environment; Feather River Coordinated Resources Management; La Union del Pueblo Entero; People for Children's Health and Environmental Justice; and Winnemem Wintu tribe.

Environmental Justice Foundation (EJF)
Web site: http://www.ejfoundation.org/
e-mail: info@ejfoundation.org

The Environmental Justice Foundation (EJF) was established in the United Kingdom in 2001 for the purpose of "investigating, exposing and resolving environmental abuses" in any and all parts of the world. Over the past decade, EJF has worked with regional and national civil rights and environmental justice organizations in 15 countries, including Cambodia, Guatemala, Guinea, and Uzbekistan. The organization has produced 17 campaign reports, 6 films, and a training manual for grassroots activities.

Environmental Justice Resource Center (EJRC)
Web site: http://www.ejrc.cau.edu/
e-mail: ejrc@cau.edu

The Environmental Justice Resource Center (EJRC) is one of the nation's oldest organizations devoted to research, technical, and legal support for communities dealing with environmental racism and related problems. The center was founded in 1994 and has long been led by Dr. Robert D. Bullard, commonly known as the father of the environmental justice movement. Much of the center's work is organized into programmatic areas, such as the Program to Facilitate Environmental Justice in Minority and Under Served Communities, the Race and Regional Equity project, the Disaster and Emergency Response program, and the National Equity and Smart Growth Initiative. The center also offers three training programs: the Brownfields Job Training Program, the Minority Worker Training Program, and the Atlanta Housing Authority Youth Apprenticeship Training Program.

Environmental Law Alliance Worldwide (ELAW)
Web site: http://www.elaw.org/
e-mail: elawus@elaw.org

The Environmental Law Alliance Worldwide (ELAW) was
formed in 1989 at a University of Oregon School of Law conference on public interest law. A group of attendees realized that the
nations from which they came all faced similar environmental
challenges and that their work could be facilitated by developing
a network of lawyers with common interests and skills. ELAW
now has more than 300 members from 60 nations who share scientific equipment and training to monitor environmental conditions, develop model laws and regulations, provide expert
testimony on pollution issues, assist in the drafting of petitions
and other court filings, share environmental and human rights
records of multinational corporations, and provide support for
colleagues in the drafting and conduct of cases.

Environmental Law and Policy Center (ELPC)
Web site: http://www.elpc.org/
e-mail: elpcinfo@elpc.org

The Environmental Law and Policy Center (ELPC) was founded
in 1993 for two major purposes: to provide the Midwest with (1)
public interest legal, economic, and scientific resources for dealing with area-wide environmental and energy issues, and (2) a
strategic regional perspective on issues such as energy, transportation, and land use and conservation. Today, ELPC works on
a large and variable group of problems, including dirty coal-
powered plants, oil refineries, farm energy, global climate
change, nuclear power, electrical grid connections, municipal
utilities, high-speed rail systems, sustainable growth, clean water, water conservation, and forest preservation. The center focuses on problems in the nine states that make up the upper
Midwest region: Illinois, Indiana, Iowa, Michigan, Minnesota,
North Dakota, Ohio, South Dakota, and Wisconsin.

Environmental Leadership Program (ELP)
1609 Connecticut Ave NW, #400
Washington, DC 20009
Telephone: (202) 332-3320
Web site: http://www.elpnet.org/

In the late 1990s, a group of young men and women interested in environmental issues in the United States met to discuss the possibility of forming an organization to train leaders like themselves for the environmental movement. The program they developed admitted its first group of 22 student leaders in December 1999. Since that time, the Environmental Leadership Program (ELP) has expanded to include three campuses: the Delaware Valley (serving New Jersey, central and eastern Pennsylvania, and Delaware), New England (Connecticut, Massachusetts, Maine, New Hampshire, Rhode Island, and Vermont), and the Southeast (Alabama, Georgia, northern Florida, North and South Carolina, and central and eastern Tennessee). Each year, ELP selects about 20 to 25 promising leaders in the environmental movement to study at each of the three regional centers. The organization provides instruction, networking opportunities, and financial support to help improve their effectiveness as environmental leaders. Many of the current leaders in the environmental justice movement are products of ELP programs, including Torri Estrada, Luis Fernandez, Shalini Gupta, Marcus Johnson, Khiet Luong, Janani Narayanan, Na'Taki Osborne, Pablo Padilla, and Swati Prakash.

Environmental Research Foundation (ERF)

Web site: http://www.rachel.org/
e-mail: info@rachel.org

The Environmental Research Foundation (ERF) is a 501(c)(3) nonprofit organization founded in 1980 on toxics and social justice issues in New Jersey. In 1986, it began publishing the journal for which it is probably best known, *Rachel's Environment and Health News* (now *Rachel's Democracy and Health News*). The goal of the journal is to show the connection among a variety of social, economic, political, environmental, and other issues that, at first glance, seem independent of each other. It is now one of the most highly respected sources for information on the effects of hazardous wastes on human health and the environment. In 2005, ERF added a sister publication, *Precaution Reporter,* which reports on current developments in the use of the precautionary principle to deal with environmental problems. The precautionary principle says that, if some policy or action is likely to cause severe or irreversible harm to the public, the burden of proof for initiating that policy or action falls on advocates for taking the action (unless there is indisputable scientific evidence to the contrary). ERF and *Rachel's News* make a clear and consistent connection between the information they provide and the goals and activities of the environmental justice movement.

Environmental Support Center (ESC)
Web site: http://www.envsc.org
e-mail: See "Contact Information" page on Web site

As its name suggests, the mission of the Environmental Support Center (ESC) is to provide administrative, organizational, and other forms of the support environmental groups of all sizes and types need to continue operating effectively and efficiently. The center was formed in 1999 following meetings among a group of local, regional, and state leaders in the environmental justice and environmental advocacy movements. One of ESC's first programs was its Training and Organizational Assistance Program, which, between 1991 and 2007, made over 3,200 small grants to more than 1,800 grassroots groups and coalitions. The center's two other major programs are its Leadership and Enhanced Assistance and its Fundraising for Sustainable Organizations programs. In addition to these programs, ESC offers technical assistance to groups through its Technology Resources program and maintains a directory of consultants, management support organizations, and other people who help groups with skills and tools.

Global Community Monitor (GCM)
Web site: http://gcmonitor.org/
e-mail: denny@gcmonitor.org

Global Community Monitor (GMC) is an international organization whose goal is to assist industrial communities in developing clean and sustainable environments in which to live and work. Its partners currently include the Louisiana Bucket Brigade, Citizens for Environmental Justice (Corpus Christi, Texas), groundWork (South Africa), and the South African Exchange Program on Environmental Justice. GMC currently operates six major campaigns: Oceans and Communities Project, Citgo Justice, Ireland: Shell to Sea, Shell Facts, Refinery Reform Campaign, and India Community Environmental Monitoring.

The organization provides a wide array of tools for use by environmental justice workers, including an air pollution manual, an introduction to wipe sampling, an introduction to soil sampling, odor analysis techniques, and odor pollution detection systems. It has also published reports on a variety of hazardous chemicals, including dioxins, PCBs, perchlorates, and polyvinyl chloride (PVC).

Greenpeace International

Web site: http://www.greenpeace.org/international/
e-mail: supporter.services@int.greenpeace.org

Greenpeace USA

Web site: http://www.greenpeace.org/usa/
e-mail: info@wdc.greenpeace.org

Greenpeace began in 1971, when a small group of environmental activists rented the fishing boat *Phyllis Cormack* to mount a protest against an underground nuclear test on the island of Amchitka off the west coast of Alaska. Over the next four decades, the organization grew to become one of the world's largest and most highly respected environmental organizations. Today, Greenpeace has offices in more than 40 countries, ranging from Spain and Portugal to Lebanon and Israel to Fiji and Papua New Guinea. The organization operates a number of major environmental campaigns dealing with issues such as climate change, protection of the oceans and ancient forests, elimination of toxic chemicals, and reduction of genetic engineering programs. Greenpeace USA has been very active in a variety of domestic environmental justice campaigns, and Greenpeace International has been at the forefront of efforts to eliminate the shipment of hazardous waste materials across national boundaries.

Honor the Earth

Web site: http://www.honorearth.org/
e-mail: info@honorearth.org

Honor the Earth was formed in 1993 by Winona LaDuke and Indigo Girls members Amy Ray and Emily Saliers to educate the general public about the many environmental issues faced by

Native Americans. The organization's board of directors consists of representatives of two national Native American groups—the Indigenous Environmental Network and the Indigenous Women's Network. Honor the Earth achieves its goal through music, the arts, and the media. The profits from its programs are used to make grants to organizations working on indigenous environmental issues.

Indian Law Resource Center (ILRC)
Web site: http://www.indianlaw.org/
e-mail: mt@indianlaw.org

Land rights, environmental protection, native sovereignty and self-government, human rights, and law reform are the major issues in which the Indian Law Resource Center (ILRC) is involved. Founded in 1978, the center is a tax-exempt organization under section 501(c)(3) of the Internal Revenue Code. It is funded entirely by grants and contributions from Indian nations and receives no other governmental support. An example of the center's work in the area of environmental justice is its 10-year effort on behalf of the Ahahnenin and Nakota tribes, whose land and water resources have been extensively contaminated by two gold mines located adjacent to the Fort Belknap Indian Reservation in Montana.

Indigenous Environmental Network (IEN)
Web site: http://www.ienearth.org/
e-mail: ien@igc.org

The Indigenous Environmental Network (IEN) was founded in 1990 to address issues of economic and environmental justice affecting indigenous peoples. The organization's mission is to strengthen the capacity of indigenous communities and tribal governments to deal with a variety of issues, such as building economically sustainable communities, protecting the health of people and the environment, and guarding sacred sites and natural resources in indigenous communities. IEN activities include the maintaining of an informational clearinghouse; organizing campaigns and direct actions; working with communities on environmental justice issues; building alliances among indigenous communities and tribes, people-of-color and ethnic organizations, women's groups, and labor and environmental organizations; and conducting local, regional, national, and international conferences and other meetings on economic and environmental justice issues.

Just Transition Alliance (JTA)
Web site: http://www.jtalliance.org/
e-mail: justtransition@sbcglobal.net

The Just Transition Alliance (JTA) is a coalition of labor, economic, and environmental justice activists; indigenous people; and working-class people of color formed in 1996. Its name suggests its mission, namely, to develop methods by which communities where working conditions are dangerous can be converted into communities that are healthy and safe with a sustainable economy. Members of the alliance include the Asian Pacific Environmental Network; the Farmworker Network for Economic and Environmental Justice; the Indigenous Environmental Network, the Northeast Environmental Justice Network; the United Steel, Paper and Forestry, Rubber, Manufacturing, Energy, Allied Industrial; Service Workers of the United Steelworkers; and the Southwest Network for Environmental and Economic Justice. As an example of its work, JTA developed an alliance among workers at the Arizona Portland Cement factory in Rillito, Arizona, the local community, and Tucsonians for a Clean Environment to deal with ongoing air pollution problems resulting from the plant's operation.

Madres del Este de Los Angeles Santa Isabel
Web site: http://clnet.ucla.edu/community/intercambios/
　　melasi/
e-mail: mothermelasi@earthlink.net

Madres del Este de Los Angeles Santa Isabel (Mothers of East Los Angeles–Santa Isabel) was formed in 1984 with the sole purpose of fighting the construction of a state prison in East Los Angeles. That battle ended successfully for the organization when the state legislature passed a bill prohibiting the construction of state prisons in Los Angeles County. After that success, the group continued to work on other environmental problems facing the community. In 1987, for example, it successfully blocked the construction of an oil pipeline that would have been laid under a local junior high school. Today, Madres del Este sponsors a number of programs focusing on topics such as water conservation, lead poison awareness, graffiti abatement, the preservation of Mono Lake, and a mentoring task force.

Military Toxics Project (MTP)
Web site: http://www.miltoxproj.org/
e-mail: mtp@miltoxproj.org

The Military Toxics Project (MTP) was started in 1989 by the National Toxics Campaign as a way of providing support and assistance to grassroots organizations in areas where military activities resulted in health and environmental risks to the community. Among the issues with which MTP has had to deal are solid and liquid wastes produced by the construction and use of chemical weapons, nuclear materials, and conventional munitions, as well as pollution and contamination resulting from base operations and closures. The organization has experienced a long series of successes in its activities. It has been instrumental, for example, in assuring that attention to environmental issues is written into legislation dealing with military base closures.

National Black Environmental Justice Network (NBEJN)
Web site: http://www.nbejn.org/
e-mail: info@nbejn.org

The National Black Environmental Justice Network (NBEJN) was formed in December 1999 at a meeting of more than 300 black grassroots economic, social, and environmental activists. The organization was originally called the Interim National Black Environmental and Economic Justice Coordinating Committee (INBEEJCC). It adopted its present name in 2001. NBEJN now consists of environmental justice activists, community organizers, researchers, lawyers, public health specialists, technical experts, and authors from 33 states and the District of Columbia who are interested in the impact of economic development, public health, and environmental hazards on black communities. The organization currently focuses its activities on four major areas: (1) safe and healthy communities; (2) sustainable development, climate justice, and clean energy production; (3) civil rights and equal protection laws and policies; and (4) international human rights protection.

National Tribal Environment Council (NTEC)
2501 Rio Grande Blvd. N.W., Suite A
Albuquerque, NM 87104
Telephone: (505) 242-2175

The National Tribal Environment Council (NTEC) is a membership organization with about 184 member tribes. It was formed in 1991, when seven tribes joined together to create an organization that provides assistance for the protection of Native American lands from environmental degradation. NTEC offers programs that allow tribes to share information about common environmental problems and to join with each other in dealing with such problems. It maintains a clearinghouse of environmental information available not only to member tribes but also to any Native American group that requests assistance. Each year, the council sponsors an annual convention dealing with an environmental issue of special concern.

Natural Resources Defense Council (NRDC)
Web site: http://www.nrdc.org/
e-mail: nrdcinfo@nrdc.org

The Natural Resources Defense Council (NRCD) is a nonprofit organization of more than 1.2 million members and a staff of lawyers, scientists, and other environmental specialists. The council is dedicated to protecting the world's natural resources and ensuring a safe and healthy environment for all people. It works through legal action, scientific research, advocacy, and public education. One of NRDC's current areas of concern is environmental justice. It provides guides, fact sheets, a photo album, and other resources related to environmental justice issues on its Web site at http://www.nrdc.org/ej/.

Service Training for Environmental Progress (STEP)
Web site: http://www.mc.vanderbilt.edu/root/vumc.php?site
    =step&doc=813
e-mail: barbara.clinton@vanderbilt.edu

Service Training for Environmental Progress (STEP) is a service of the Vanderbilt University Medical Center (VUMC) in Nashville. STEP originated in 1981 as the Student Environmental

Health Project, which itself evolved out of the university's Occupational Health Project. The organization's purpose is to provide assistance to community-based environmental organizations in low-income communities. STEP places students in communities to assist local leaders in educating residents about health issues. The program also has provisions by which communities can sponsor Vanderbilt interns to work on environmental issues within a community. This program additionally allows the community to take advantage of a variety of university resources.

Sierra Club
Web site: http://www.sierraclub.org/
e-mail: information@sierraclub.org

The Sierra Club was formed in 1892 to defeat a proposed reduction in the size of Yosemite National Park. Over the years, it has become one of the leading conservation groups in the United States, with programs in a diverse range of fields, such as urban sprawl, clean water, forest protection, global warming, human rights, and global population. It currently sponsors environmental justice projects in eight parts of the United States, including Detroit; Appalachia; New Orleans; Flagstaff, Arizona; and El Paso, Texas. Its Web page provides information about these projects and about environmental justice news and events. The club also publishes a newsletter, *EJ Activist*.

Student Environmental Action Coalition (SEAC)
Web site: http://www.seac.org/
e-mail: seac@seac.org

The Student Environmental Action Coalition (SEAC) grew out of a national meeting, Threshold, held at the University of North Carolina in Chapel Hill on October 27–29, 1989. More than 1,700 students from 43 states and over 225 colleges and universities attended the meeting. The organization has experienced varying degrees of success over the past two decades, partly because of inadequate funding and partly because of changes in leadership. It currently is focusing on two major campaigns, one dealing with global climate change and the other with the health risks of tampons. It is also conducting three major initiatives, one supporting the Coalition of Immokalee Workers, a group of Florida workers in low-wage jobs throughout the state; one on Militarism and the Environment, dealing with the connection between military action and environmental degradation; and one on Mountain Justice, focusing on the environmental damage caused by surface coal mining.

United States Conference of Catholic Bishops' Environmental Justice Program: Caring for God's Creation (USCCB-EJP)
Web site: http://www.usccb.org/sdwp/ejp/
e-mail: sdwpmail@usccb.org

The United States Conference of Catholic Bishops' Environmental Justice Program (USCCB-EJP) was initiated in 1993 to educate Catholics and motivate them to have a greater appreciation of God's creations, as well as to encourage parishes and dioceses to become active in programs to promote environmental justice. Today, USCCB-EJP supports a program consisting primarily of four elements: scholarly activities; leadership development; public policy and advocacy; and special projects, such as children's environmental health, youth leadership development, and regional grants.

State, Regional, and Local

Alaska Community Action on Toxics (ACAT)
Web site: http://www.akaction.org/
e-mail: info@AKAction.net

Alaska Community Action on Toxics (ACAT) was organized in 1997 to deal with environmental justice issues in Alaska and other regions. It currently has four major programs focusing on military toxics and health, special issues related to the presence of contaminants and health in northern regions, right-to-know issues involving pesticides, and water quality protection. ACAT provides services associated with geographic information and computer mapping, investigative research, advocacy, and training. The organization has produced a number of reports, newsletters, and fact sheets, the most recent of which are the spring 2007 "Health and Environment Bulletin for Health Care Providers" and a newsletter, *Reproductive Health and the Environment*.

Alternatives for Community and Environment (ACE)
Web site: http://www.ace-ej.org/
e-mail: info@ace-ej.org

Alternatives for Community and Environment (ACE) was founded in 1993 to assist residents of Boston's predominantly black Roxbury neighborhood in dealing with its environmental problems. Since that time, the organization has expanded its mission by partnering with more than 40 groups in the greater Boston area and the cities of Lowell, Lawrence, and New Bedford, Massachusetts. Some of the organization's accomplishments include the cleanup of large amounts of asbestos and toxic-laden dirt in Roxbury, promotion of the adoption of pollution-free buses by local transportation authorities, successful opposition to a diesel-fueled power plant in the Chelsea neighborhood of Boston, and construction of an air monitoring station in the Dudley Square neighborhood of Boston.

Center for Community Action and Environmental Justice (CCAEJ)
Web site: http://www.ccaej.org/
e-mail: admin@ccaej.org

The Center for Community Action and Environmental Justice (CCAEJ) is a 501(c)(3) nonprofit organization located in Riverside, California. The center's goal is to bring groups of people together to work on problems of common interest, a goal it achieves primarily by partnering with other organizations to provide information and publications; outreach, referral, and network development; and training and leadership programs. The organization's primary fields of interest are air quality and the movement of goods; the desert protection campaign; the perchlorate campaign; the community empowerment campaign; and the local voices, local votes campaign.

Communities for a Better Environment (CBE)
1440 Broadway, Suite 701
Oakland, CA 94612
Telephone: (510) 302-0430
Web site: http://www.cbecal.org
5610 Pacific Boulevard, Suite 203

Huntington Park, CA 90255
Telephone: (323) 826-9771

Communities for a Better Environment (CBE) works with low-income Latinos, African Americans, and other minority groups to deal with problems of environmental pollution. The group provides legal assistance, scientific and policy research and advice, and aid in organizing the community. It operates on the assumption that local populations are best able to understand and deal with the environmental problems affecting their communities.

Deep South Center for Environmental Justice (DSCEJ)
Web site: http://www.dscej.org/
e-mail: dscej@aol.com

The Deep South Center for Environmental Justice (DSCEJ) was founded in 1993 to provide a means by which universities and communities could work together to deal with issues of environmental justice. The center is located at Dillard University in New Orleans, in a region where many communities are severely affected by environmental hazards. One of DSCEJ's major goals is to develop minority leadership in the areas of social, economic, and environmental justice for communities along the Mississippi River corridor. It uses a model called "communiversity," which makes use of the special knowledge and skills of community leaders and academicians. Since 2005, the center has also focused its resources and skills in helping the New Orleans area recover from and rebuild after the devastating Hurricane Katrina of that year.

Detroiters Working for Environmental Justice (DWEJ)
Web site: http://www.dwej.org/
e-mail: See "Contact Us" page on Web site

Detroiters Working for Environmental Justice (DWEJ) was organized in 1994 to provide assistance to minority and low-income people in the Detroit area in dealing with problems of environmental racism. The organization currently operates a half-dozen major programs on topics such as hazard awareness training and lead dangers in the home, is involved in a state campaign for environmental justice, and runs a program for monitoring pollution by youth in the community.

Greenaction
Web site: http://www.greenaction.org/
é-mail: greenaction@greenaction.org

Greenaction was formed in the late 1990s when Bradley Angel, then a member of Greenpeace, felt that the organization was not doing enough to deal with issues of environmental justice. He organized a new group whose first action was to support a protest by five Native American tribes against the construction of a nuclear waste dump in California's Ward Valley. It was successful in convincing Gov. Gray Davis to issue an order prohibiting further work on the dump. Currently, Greenaction supports an expanded program that includes a campaign to deal with pollution issues in San Francisco's Bayview-Hunters Point neighborhood, a protest against the use of certain types of incinerators for solid waste disposal, a campaign against pollution produced by Pacific Steel's casting operations, and two environmental justice programs—one focusing on the environmental health, justice, and sovereignty of indigenous lands, and one on environmental health and justice issues in California's San Joaquin Valley.

KATRINA ORGANIZATIONS
When Hurricane Katrina struck the New Orleans area in August 2005, minority and low-income communities in the region experienced a significant deterioration in their environment, an issue with which they had already been dealing for many years. When the outside assistance needed for recovery from hurricane damage failed to materialize, local residents formed a number of grassroots organizations to deal with their problems using their own resources. As of 2008, more than two dozen such organizations are still in operation. Their names and links to their Web sites can be found on the Internet at http://www.cwswork shop.org/katrinareader/view/org_listing. The following organization is typical of the kinds of groups found on that list.

New Orleans Workers' Center for Racial Justice (NOWCRJ)
803 Baronne Street
New Orleans, LA 70113
Telephone: (504) 309-5165
Web site: http://www.neworleansworkerjustice.org

The New Orleans Workers' Center for Racial Justice (NOWCRJ) was organized in August 2006 by a coalition of law students from about a dozen organizations, including the Advancement Project, Common Ground, Gillis Long Poverty Law Center, Hope House, Latino Health Outreach Project, Louisiana Environmental Action Network, National Immigration Law Center, New Orleans Students United for Worker Justice, and Safe Streets/Strong Communities. The original purpose of the coalition was to collect information for a publication, *And Injustice for All: Workers' Lives in the Reconstruction of New Orleans*. Since that time, NOWCRJ has continued to function as a grassroots organization working for racial justice across races and occupations. Three of its current projects are the Alliance of Guest Workers for Dignity, working for the rights of foreign workers who have come to New Orleans since the hurricane; the Congress of Day Laborers, which focuses on problems of workers who depend on day work to survive; and the Homeless Worker Organizing Project, which offers aid and resources to the estimated 12,000 homeless residents of the city.

Liberty Hill Foundation–Environmental Justice Fund (EJF)
Web site: http://libertyhill.org/donor/environment.html
e-mail: info@libertyhill.org

Liberty Hill Foundation is a Los Angeles–based organization that makes grants to combat poverty and injustice and to transform the city into a community that "promises safety, equality and opportunity for everyone who lives here." Its activities are divided into four major areas: providing training and networking for grassroots leaders in the community, making grants, providing programs and services, and helping to shape the debate on key issues in the community. Between 1996 and 2008, the foundation made grants of more than $2.4 million to environmental justice groups through its Environmental Justice Fund (EJF). Examples of the groups funded include Amigos de los Rios, the California Environmental Rights Alliance, the Coalition for Economic Survival, Del Amo Action Committee, and Los Angeles Alliance for a New Economy.

Literacy for Environmental Justice (LEJ)
Web site: http://www.lejyouth.org/
e-mail: ej@lejyouth.org

Literacy for Environmental Justice (LEJ) is an organization based in the Bayview-Hunters Point neighborhood of San

Francisco whose mission is to promote an understanding of environmental justice among young men and women of the community. Since its founding in 1998, LEJ has involved more than 10,000 young people in programs designed to promote its goals and objectives. Its current efforts are organized into six major programmatic areas: Environmental Health and Justice Education, Heron's Head Park Education, the Living Classroom Program, Youth Envision Program, Slough Youth Program, and Youth for Environmental Health Program. The organization's educational programs include an initial project planning meeting with teachers and students, a series of lessons that define and explain the issue to be studied, field research and/or community investigations, an action project, and a final evaluation.

Little Village Environmental Justice Organization (LVEJO)
Web site: http://www.lvejo.org/
e-mail: info@lvejo.org

Little Village Environmental Justice Organization (LVEJO) is a neighborhood grassroots organization of people living in the Little Village neighborhood of Chicago that works for economic, environmental, and social justice. The organization's current campaigns focus on the cleanup of the Celotex section of Little Village, the promotion of clean power sources for the area, the mapping of the community and its resources, the development of clean and safe Latino schools, the expansion of park and recreational areas in the neighborhood, and the promotion of urban agricultural projects. The organization publishes a regular newsletter about its projects and accomplishments, *El Cilantro.*

Louisiana Environmental Action Network (LEAN)
Web site: http://leanweb.org/
e-mail: contact@leanweb.org

The Louisiana Environmental Action Network (LEAN) consists of more than 100 grassroots organizations and more than 1,700 individuals working to deal with environmental issues caused by one of the most hazardous collections of industrial facilities anywhere in the United States. The network works with individual groups that are attempting to deal with local conditions by providing expert advice, media resources, leadership training, and legal advice. The organization also conducts the annual

LEAN People's Conference providing members with support, solidarity, and education. Recent network campaigns have focused on the environmental effects of hurricanes Rita and Katrina, management of hurricane debris, climate change issues, and threats posed to cypress resources as a result of cypress mulch collection.

New York City Environmental Justice Alliance (NYCEJA)
Web site: http://www.nyceja.org/
e-mail: generalinfo@nyceja.org

The New York City Environmental Justice Alliance (NYCEJA) is a 501(c)(3) corporation formed in 1991 to bring together the efforts of a number of environmental justice groups working in New York City. Today NYCEJA consists of 15 local grassroots organizations such as El Puente, Lower Washington Heights Neighborhood Association, Magnolia Tree Earth Center, Morningside Heights/West Harlem Sanitation Coalition, Nos Quedamos, and United Puerto Ricans' Organization of Sunset Park (UPROSE). NYCEJA action is currently organized around four major campaigns: Apollo and Energy Advocacy, which promotes sustainable environmental and economic development; Brownfield Redevelopment and Advocacy, which involves identifying brownfields in the community and working to develop safe and useful alternatives for the land; Open Space Equity Campaign, which works to increase the amount of open space in neighborhoods of poor and minority people; and Solid Waste Management Advocacy, which deals with inequities in the siting of solid waste disposal sites. NYCEJA publishes newsletters for each of these campaigns and has produced two major reports on environmental justice in New York City: *New York Urban Environment: Underfunded and Underserved* and *Reaching for Zero: A Citizens Plan for Zero Waste in New York City.*

SouthWest Organizing Project (SWOP)
Web site: http://www.swop.net/
e-mail: swop@swop.net

The SouthWest Organizing Project (SWOP) was founded in 1981 as a statewide, multiracial, multi-issue, grassroots membership organization in New Mexico. One of its guiding principles is that people will live peacefully with each other only with the elimination of racism, sexism, and age and social class discrimination.

The project is working for the self-determination of all peoples and for social, environmental, and economic justice at home and abroad. Among its activities are the registration of voters, primarily in Hispanic American and African American communities; efforts to promote community control of housing, zoning, and local economic development; the promotion of health and safety concerns of workers; and the development of educational materials. Its current campaigns include Intel: What Are We Breathing, an effort to hold Intel accountable for damage it has caused to the environment and economy; Youth, a program for the development of young leaders in the community; Bucket Brigade, a community air pollution monitoring program; and Pajarito Mesa, a campaign to assist a community facing especially difficult economic and environmental issues.

Southwest Research and Information Center (SRIC)
Web site: http://www.sric.org/
e-mail: info@sric.org

The Southwest Research and Information Center (SRIC) is one of the oldest environmental justice groups in the United States. It was formed in 1971 to provide citizens of the southwestern United States with the information they needed to understand how their lives were being impacted by a variety of operations, such as uranium mining and nuclear waste disposal, and how they could most effectively deal with those problems. The organization's mission has not changed significantly in the last four decades. Currently, its activities are organized into five major programs: uranium impact assessment, community development and economics, energy and natural resources, environmental information and education, and nuclear waste safety. SRIC publishes a quarterly journal, *Voices from the Earth.*

West Harlem Environmental Action (WE ACT)
Web site: http://www.weact.org/
e-mail: Peggy@weact.org

One of the oldest, best-known, and most prestigious environmental justice organizations is West Harlem Environmental Action (WE ACT). It was formed in 1988 as a means through which community members could begin to take action against the operation of the notorious North River Sewage Treatment Plant in West Harlem and the proposed construction of a diesel bus depot in northern Manhattan. Today, WE ACT is a community-based, non-profit organization working to inform, educate, train, and mobilize the predominantly African American and Latino residents of northern Manhattan. The organization deals with any issue that affects the quality of life in the area, including air and water pollution, housing problems, toxic materials, land use and open space, waterfront development and usage, sanitation, transportation, historic preservation, regulatory enforcement, and citizen participation in public policy making.

Examples of the activities in which WE ACT has been engaged are health surveys, a youth program to build awareness of environmental issues, a lead poisoning prevention program, a neighborhood environmental center, community forums, and a community environmental resources library.

8

Resources

Environmental justice is still a relatively young field. Insufficient time has passed to permit the development of an extensive literature base about the subject, as is now the case in other fields of social injustice or environmentalism. This chapter lists some of the most important works currently available on environmental justice, environmental racism, and related topics. The chapter is divided into four major sections: books, articles, reports, and Internet sources.

The resources in this chapter deal with all aspects of the environmental justice movement, including topics such as the history of the movement, research evidence dealing with the extent and characteristics of environmental inequities, methodological issues in the study of environmental inequities, grassroots and other types of efforts to deal with environmental inequities, the relationship of the environmental justice movement with sustainable development and climate change, case studies in the battle against environmental inequities, international aspects of the environmental justice movement, responses to environmental inequities by governmental agencies and businesses, legal issues related to the environmental justice movement, and regulatory issues in dealing with environmental inequities.

Books

Adamson, Joni, Mei Mei Evans, and Rachel Stein, eds. *The Environmental Justice Reader: Politics, Poetics, & Pedagogy.* Tucson: University of Arizona Press, 2002.

245

A fascinating book that claims to be the first text "to link political studies, literary analysis, and teaching strategies." Accordingly, the book includes a wide-ranging collection of articles by scholars and workers in the field on some general principles of environmental justice as well as a number of interesting and instructive case studies from the United States and other parts of the world.

Agyeman, Julian. *Sustainable Communities and the Challenge of Environmental Justice.* **New York: NYU Press, 2005.**

The author begins by providing a general introduction to environmental justice and then explains the special problems faced in developing sustainable communities where environmental inequities exist. He then discusses the concept of "just sustainability" in both theory and practice, concluding with some specific recommendations for finding ways of providing communities with both a sustainable basis and environmental justice.

Allen, Barbara L. *Uneasy Alchemy: Citizens and Experts in Louisiana's Chemical Corridor Disputes.* **Cambridge, MA: MIT Press, 2003.**

The author analyzes a number of events in which social and environmental activists in Louisiana's Cancer Alley confronted chemical and petroleum industries in the area over the pollution they were producing. She explores the factors that led to the considerable number of successes achieved by citizen groups and finds that a critical factor in success was the establishment of alliances among grassroots activists and experts in the field of environmental health and related areas.

Anand, Ruchi. *International Environmental Justice: A North-South Dimension.* **Aldershot, UK: Ashgate Publishing, 2005.**

This book discusses environmental inequities in three general areas: climate change, ozone politics, and hazardous waste disposal. The author points out some major themes in international environmental inequities that parallel those found in domestic cases in the United States.

Bowen, William M. *Environmental Justice through Research-Based Decision-Making.* **London: Routledge Press, 2001.**

This scholarly treatise presents a detailed analysis of the research technology available for the study of environmental inequity problems. Part I deals with general methodological issues, such as the definition of terms and concepts used in the field, research methods available for the study of environmental inequity issues, and the design and validation of research studies. Part II reviews existing research on environmental inequities and suggests some general principles for the design of future studies. Part III outlines some lessons to be learned from research on environmental inequities for administrative decision making.

Bryant, Bunyan. *Environmental Justice: Issues, Policies, and Solutions.* **Washington, DC: Island Press, 1995.**

This collection of articles covers a broad range of topics, such as "Health-Based Standards: What Role in Environmental Justice?" "Environmental Justice Centers: A Response to Inequity," "Environmentalists and Environmental Justice Policy," "Residential Segregation and Urban Quality of Life," "The Net Impact of Environmental Protection on Jobs and the Economy," "Towards a New Industrial Policy," "Environmental Equity and Economic Policy: Expanding the Agenda of Reform," "Indigenous Nations: Summary of Sovereignty and Its Implications for Environmental Protection," "Sustainable Agriculture Embedded in a Global Sustainable Future: Agriculture in the United States and Cuba," and "Rethinking International Environmental Policy in the Late Twentieth Century."

Bryant, Bunyan, and Paul Mohai, eds. *Race and the Incidence of Environmental Hazards.* **Boulder, CO: Westview Press, 1992.**

This book is one of the classic references in the field of environmental justice. Its 14 articles are divided between discussions of specific examples of environmental racism ("Environmentalism and Civil Rights in Sumter County, Alabama," "Invitation to Poison? Detroit Minorities and Toxic Fish Consumption from the Detroit River," and "Uranium Production and Its Effects on

Navajo Communities along the Rio Puerco in Western New Mexico") and more general discussions of fundamental issues in environmental justice ("Toxic Waste and Race in the United States," "Can the Environmental Movement Attract and Maintain the Support of Minorities?" and "Environmental Black Mail in Minority Communities").

Bullard, Robert D., ed. *Confronting Environmental Racism: Voices from the Grassroots.* **Boston: South End Press, 1993.**

Some of the 12 chapters in this anthology describe specific instances of environmental racism, including discussions as to how individuals and organizations dealt with those instances. Examples of these cases are the chapters "Race and Waste in Two Virginia Communities," "Environmental Politics in Alabama's Blackbelt," and "Sustainable Development at Gannados del Valle." Other chapters consider broader issues in environmental justice, such as Bullard's introductory chapter, "Anatomy of Environmental Racism and the Environmental Justice Movement," and later chapters titled "Environmentalism and the Politics of Inclusion" and "Coping with Industrial Exploitation."

Bullard, Robert D. *Dumping in Dixie: Race, Class, and Environmental Quality,* **3rd ed. Boulder, CO: Westview Press, 2000.**

This book is one of the seminal works in the environmental justice movement. It arose out of Bullard's research in Houston on the spatial distribution of municipal solid waste disposal sites. Bullard then extended his research to other areas in which environmental racism appeared to exist, specifically, West Dallas, Texas; Institute, West Virginia; Alsen, Louisiana; and Emelle–Sumter County, Alabama. The book reports the results of his research in these areas in addition to providing a general analysis of the nature of environmental racism and the ways in which it can be combated.

Bullard, Robert D., ed. *Growing Smarter: Achieving Livable Communities, Environmental Justice, and Regional Equity.* **Cambridge, MA: MIT Press, 2007.**

The contributors to this book, a group of urban planners, sociologists, economists, educators, lawyers, health professionals, and environmentalists, discuss the concept of "smart growth" within the context of environmental racism and environmental inequities. Smart growth is a movement that emphasizes the development of holistic communities on a pedestrian scale that value diversity and mixed land use. The authors point out how the smart growth movement has largely ignored problems related to environmental justice and suggest ways in which this deficiency can be overcome in the future.

Bullard, Robert D., ed. *The Quest for Environmental Justice: Human Rights and the Politics of Pollution.* **San Francisco: Sierra Club Books, 2005.**

This book is a follow-up to the author's highly regarded *Unequal Protection: Environmental Justice and Communities of Color* (see below) that updates the status of the environmental justice movement in the 21st century. It includes a number of case studies of the ongoing problem of environmental racism in communities such as the Cancer Alley region of Louisiana, metropolitan Los Angeles, the New Jersey waterfront, and the island of Vieques, Puerto Rico. The last chapter focuses on international aspects of the environmental inequities problem.

Bullard, Robert D., ed. *Unequal Protection: Environmental Justice and Communities of Color.* **San Francisco: Sierra Club Books, 1994.**

The essays that make up *Unequal Protection* are divided into three parts. The first part covers the historical background of the environmental justice movement. It includes a review of two early cases that helped define the movement, at Indian Creek, Alabama, and Warren County, North Carolina. Part II describes cases that have arisen in "sacrifice zones," areas where high concentrations of industrial pollution are found. Among the examples discussed in this section are a Superfund site in Texarkana, Texas, a lead smelter in Dallas, and the area of the Mississippi River Valley known as Cancer Alley. Part III of the book focuses on the alliances that have developed between mainstream environmentalists and grassroots organizations interested specifically in issues of environmental justice. Among the organizations discussed are the SouthWest Organizing Project, Concerned Citizens of South Central Los Angeles, and Mothers of East Los Angeles.

Bullard, Robert D., and Dana A. Alston, eds. *We Speak for Ourselves: Social Justice, Race, and Environment.* **Washington, DC: Panos Institute, 1990.**

Most of the 13 articles in this book describe specific instances in which activists have worked to deal with environmental issues in communities of color. Other articles discuss the issues involved in bringing together people of color and people from low-income groups with more traditional (and, usually, mostly white) environmental groups. The final chapter describes international aspects of the environmental justice issue.

Bullard, Robert D., and Glenn S. Johnson, eds. *Just Transportation: Dismantling Race and Class Barriers to Mobility.* **Gabriola Island, BC: New Society Publishers, 1997.**

The essays in this book by environmental activists, transportation experts, and lawyers point out how institutionalized racism affects the way public transportation systems are designed and built, resulting in unequal health and environmental damage to communities with large low-income and minority populations. The book not only discusses general principles but also provides specific examples of cases in which transportation decisions have had unequal effects on different kinds of local communities.

Camacho, David E., ed. *Environmental Injustices, Political Struggles: Race, Class, and the Environment.* **Durham, NC: Duke University Press, 1998.**

The essays that make up this book are divided into four sections, the first of which provides a general theoretical framework for the environmental justice movement. The second section discusses specific examples of environmental inequity, whereas the third section reviews some methods that have been developed for dealing with such inequities. The final section, "Environmental Justice," contains essays with mixed foci.

Caruthers, David V. *Environmental Justice in Latin America: Problems, Promise, and Practice.* **Cambridge, MA: MIT Press, 2008.**

While the environmental justice movement in the United States and, more recently, Europe and Africa have been studied in some detail, there has been relatively little attention given to the problems of environmental inequity in Latin America. This book attempts to remedy that deficiency by outlining some theoretical concepts with unique application to Latin America as well as an extensive review of some instructive case studies from the region.

Checker, Melissa. *Polluted Promises: Environmental Racism and the Search for Justice in a Southern Town.* **New York: NYU Press, 2005.**

The author tells the story of yet one more community affected by environmental inequities (the Hyde Park neighborhood of Augusta, Georgia), reviews the health problems endemic in the area, and discusses the activities of local citizens fighting for environmental justice.

Cole, Luke W., and Sheila R. Foster, eds. *From the Ground Up: Environmental Racism and the Rise of the Environmental Justice Movement.* **New York: NYU Press, 2000.**

The authors draw from their own experiences as activists in the environmental justice movement as well as from specific case studies of communities confronted by environmental inequities to review the history of the environmental justice movement and to analyze methods that have proved to be and not to be successful in dealing with environmental racism. Reviews of the book praise its ability to draw some important general conclusions from the specific cases presented.

Faber, Daniel Richard, ed. *The Struggle for Ecological Democracy: Environmental Justice Movements in the United States.* **New York: Guilford Press, 1998.**

The dozen essays in this book discuss a number of issues related to the environmental justice movement, including specific case studies, such as Hispanic use of land for grazing purposes in the Southwest, environmental politics in the upper Rio Grande, and the Earth First! movement in northern California, as well as more general analyses of issues such as the political ecology of American capitalism, the use of epidemiology in environmental justice studies, and the effects of capitalist production systems on the environment.

Foreman, Christopher H. *The Promise and Peril of Environmental Justice.* Washington, DC: Brookings Institution Press, 1998.

Although acknowledging a number of victories as a result of the environmental justice movement, the author claims that many challenges still face the movement, including a tendency to accept one-sided approaches to dealing with problems of environmental inequities and an unwillingness to accept necessary compromises. For this reason, he argues, the movement may be avoiding or ignoring other critical elements related to social and health problems confronting low-income and minority communities.

Gerrard, Michael B., and Sheila R. Foster, eds. *The Law of Environmental Justice,* 2nd ed. Chicago: American Bar Association, 2008.

The 21 papers in this book discuss legal theory and practice related to environmental justice and include topics such as federal and state regulations and programs, Native American law, access to information, impact and risk assessment, challenges to permits for new and existing facilities, brownfields, and residential and workplace exposure.

Gibbs, Lois Marie. *Love Canal: The Story Continues.* Gabriola Island, BC: New Society Publishers, 1998.

Although not commonly thought of as part of the environmental justice movement, Gibbs's campaign at the Love Canal site in New York State where she lived remains one of the signature events in the early history of the movement to protect communities against the dangers of hazardous waste dumps. This book retells that story and explains how the Love Canal battle led to the development of the Center for Health, Environment and Justice, which remains one of the premier organizations in the environmental justice movement.

Girdner, Eddie J., and Jack Smith. *Killing Me Softly: Toxic Waste, Corporate Profit, and the Struggle for Environmental Justice.* New York: Monthly Review Press, 2002.

The authors explore the subject of environmental justice from the standpoint of the growth of the industries whose products are responsible for environmental inequities. They demonstrate the economic forces that have led to the vastly increased problem of hazardous waste disposal and the mushrooming health risks this trend has produced. They focus in particular on the struggle of people in Mercer County, Missouri, against plans by Amoco Waste-Tech to build a large toxic waste landfill in the county and the success achieved by those people.

Glazer, Phyllis, Roy Flukinger, Eugene Hargrove, and Marvin Legator (photographs by Tammy Cromer-Campbell). *Fruit of the Orchard: Environmental Justice in East Texas.* **Denton: University of North Texas Press, 2006.**

Through words and photographs, this book tells the story of Winona, Texas, a small town of 500 people living downwind of a toxic waste injection well facility built in 1982 by American Ecology Environmental Services (AEES). As the result of efforts by a local environmental justice group, Mothers Organized to Stop Environmental Sins (MOSES), AEES closed the well in 1997, but not before adults and children in Winona had developed an unusually high number of health problems allegedly because of chemicals deposited in the well.

Goldman, Benjamin A. *The Truth about Where You Live.* **New York: Random House, 1992.**

Under the Freedom of Information Act, the author and his colleagues obtained voluminous data about a variety of environmental hazards to which American citizens are exposed and assessed the degree of exposure to such hazards by residents in every county and parish in the nation. The result is a book consisting of hundreds of maps and tables describing the extent to which various regions of the nation are at risk for health problems, such as heart disease, birth defects, all forms of cancers, infectious diseases, industrial toxins, and nonindustrial pollution. Chapter 6, "Environmental Justice," focuses on the disproportionate extent to which minorities and poor communities are exposed to such hazards. The book is a critical reference for those interested in studying the factual data upon which claims of environmental inequities are often based.

Grijalva, James M. *Closing the Circle: Environmental Justice in Indian Country.* **Durham, NC: Carolina Academic Press, 2008.**

The author discusses special problems created by the confluence of federal and tribal laws in dealing with environmental inequities among Indian tribes. The book analyzes air, water, and land pollution issues within the context of Indian traditions and tribal values.

Hall, Bob, ed. *Environmental Politics: Lessons from the Grassroots.* **Durham, NC: Institute for Southern Studies, 1988.**

Environmental Politics consists of 11 essays relating important case studies in the early years of the environmental justice movement. Two of the most interesting took place in Warren County, North Carolina, often thought of as the beginning of the environmental justice movement, and in Robeson County, North Carolina, the infamous case of "the river that would not die."

Harding, Andrew, ed. *Access to Environmental Justice: A Comparative Study.* **Boston: Martinus Nijhoff Publishers, 2007.**

This book reports the results of a study of 11 countries widely distributed throughout the world to determine the extent to which ordinary citizens can and do have influence over the environments in which they live, using both legal and extra-legal means. The study finds that the problems faced by communities, the methods used to deal with those problems, and the relative success in resolving the problems vary widely from place to place.

Hofrichter, Richard, ed. *Reclaiming the Environmental Debate: The Politics of Health in a Toxic Culture.* **Cambridge, MA: MIT Press, 2000.**

The essays that make up this anthology describe the general social, economic, and political conditions in the United States that make possible the environmental inequities against which the environmental justice movement struggles. The authors point out that society's unquestioned acceptance of the production of hazardous wastes, economic blight, substandard housing, chronic stress, exploitative working conditions, and dangerous technologies cumulatively provide a context in which health risks for certain groups of people are an inevitable result.

Hofrichter, Richard, ed. *Toxic Struggles: The Theory and Practice of Environmental Justice.* **Salt Lake City: University of Utah Press, 2002.**

The editor of this collection of essays explains that they are written "from a perspective that places environmental issues in a larger context of struggles for social change, locally and globally." The essays are divided into two sections, the first of which provides a theoretical perspective on the context of environmental justice. They consider topics such as "Capitalism and the Crisis of Environmentalism," "Creating a Culture of Destruction: Gender, Militarism, and the Environment," and "Feminism and Ecology." Part II of the book deals with practical issues and includes chapters such as "Blue-Collar Women and Toxic-Waste Protests: The Process of Politicization," "Ecofeminism and Grass-Roots Environmentalism in the United States," and "Farm Workers at Risk."

Legg, Janine. *Building Healthy Communities in Environmental Justice Areas.* **Charleston, SC: BookSurge Publishing, 2006.**

The author covers a wide range of topics related to health issues in communities affected by environmental inequities. She begins with a general introduction to the subject of environmental justice and explains how environmental inequities arise in some communities and not in others. She then describes a number of research tools that can be used to assess the presence and extent of health problems related to environmental factors. She concludes by discussing the role of health policy planning in the prevention of health problems in communities at risk from environmental inequities. She concludes with a discussion of a research study conducted on environmental inequities in a nine-county region of Pennsylvania.

Lerner, Steve. *Diamond: A Struggle for Environmental Justice in Louisiana's Chemical Corridor.* **Cambridge, MA: MIT Press, 2006.**

The author tells the history of the battle between the Concerned Citizens of Norco (CCN) and the Shell Oil Company over environmental inequities in Louisiana's notorious Cancer Alley. Norco is located between a chemical plant owned by Shell and an oil refinery owned by one of its subsidiaries, Motiva. CCN eventually won its battle with Shell, which made a number of adjustments in its operations to reduce its impact on the Diamond neighborhood of Norco and eventually paid to relocate many citizens of the town.

Lucas, Karen, ed. *Running on Empty: Transport, Social Exclusion and Environmental Justice.* **Bristol, UK: Policy Press, 2004.**

This book explores a topic that is often ignored in discussions of environmental justice, the unequal benefits and costs of the siting of transportation systems. Contributors to the volume discuss past failure of planners to take into consideration these benefits and costs, discuss a number of case studies that illustrate the problem of environmental inequities in transportation planning, and suggest new options being developed to overcome this deficiency.

Malone, Linda A., and Scott Pasternack. *Defending the Environment: Civil Society Strategies to Enforce International Environmental Law.* **Washington, DC: Island Press, 2006.**

The authors provide an analysis of some international environmental inequity issues (such as transnational shipments of hazardous wastes), with a discussion of legal remedies that have been and can be used to enforce existing international and national laws prohibiting such practices.

McDonald, David A., ed. *Environmental Justice in South Africa.* **Athens: Ohio University Press, 2002.**

The papers in this volume provide a comprehensive review of the emergence and evolution of environmental justice in South Africa and consider some unique characteristics and problems related to environmental justice in the nation.

McGurty, Eileen. *Transforming Environmentalism: Warren County, PCBs, and the Origins of Environmental Justice.* **New Brunswick, NJ: Rutgers University Press, 2007.**

The author presents a detailed and intensive analysis of an event that has been called the defining moment in the environmental justice movement, a protest that occurred in Warren County,

North Carolina, in 1982. She argues that the event transformed the nature of the environmentalist movement in the United States.

Pellow, David Naguib. *Garbage Wars: The Struggle for Environmental Justice in Chicago.* **Cambridge, MA: MIT Press, 2004.**

The author points out that Chicago has more landfills per square mile than any other American city, explaining perhaps its long history of controversy over the siting of garbage dumps and the deleterious effects they have on local communities. He traces the development of this ongoing problem to its origins in the 1880s and explores a number of current controversies in Chicago neighborhoods.

Pellow, David Naguib. *Resisting Global Toxics: Transnational Movements for Environmental Justice.* **Cambridge, MA: MIT Press, 2007.**

The author explores the increasing trade and dumping of hazardous wastes produced in developed nations and delivered to poorer, developing nations. He also examines the increasingly sophisticated responses by poor and ethnic communities to this practice in countries throughout the world. He particularly focuses on the wastes associated with pesticide production and with the manufacture of computer components, such as integrated chips.

Pellow, David Naguib, and Robert J. Brulle, eds. *Power, Justice, and the Environment: A Critical Appraisal of the Environmental Justice Movement.* **Cambridge, MA: MIT Press, 2005.**

The premise of this book is that, although the environmental justice movement has experienced considerable success in the last three decades, no critical analysis has yet been made of the movement's effectiveness. This book attempts to do so, focusing first on the development of the movement itself, especially in comparison to the civil rights movement. It then examines the possibility of new approaches to environmental justice activism and alternative methods of community organization. Finally, it explores the spread of the environmental justice movement to other countries, particularly with regard to the transnational trade in hazardous materials.

Pellow, David Naguib, and Lisa Park. *The Silicon Valley of Dreams: Environmental Injustice, Immigrant Workers, and the High-Tech Global Economy.* **New York: NYU Press, 2002.**

The authors explore the environmental problems created by the huge electronics industry in Silicon Valley, California, and demonstrate its disproportionate health effects on those who work in the industry, especially poor immigrants whose communities bear the brunt of the industry's waste products.

Rechtschaffen, Clifford, and Eileen Gauna. *Environmental Justice: Law, Policy, and Regulation.* **Durham, NC: Carolina Academic Press, 2002.**

This textbook is designed for courses in environmental justice and related topics. It makes use of materials from case studies, empirical research, agency actions, and laws and regulations and documents from sociology, political science, risk assessment, and environmental and civil rights laws.

Rhodes, Edwardo Lao. *Environmental Justice in America: A New Paradigm.* **Bloomington: Indiana University Press, 2005.**

The author provides an interesting review of the history of the environmental justice movement, examines the reasons that minority groups and low-income people have not generally been involved in the environmental movement, and suggests some new ways of analyzing and attacking problems of environmental inequity.

Roberts, J. Timmons, and Melissa M. Toffolon-Weiss. *Chronicles from the Environmental Justice Frontline.* **Cambridge, UK: Cambridge University Press, 2001.**

This book focuses on battles for environmental justice in Louisiana, providing first a general introduction to the history of those struggles and then a review of four specific cases in which communities challenged industry over the health and environmental damage they were causing.

Sandler, Ronald, and Phaedra C. Pezzullo. *Environmental Justice and Environmentalism: The Social Justice Challenge to the Environmental Movement.* **Cambridge, MA: MIT Press, 2007.**

Most books tend to look at environmental justice from the standpoint of human rights. This book raises issues about the movement from the standpoint of environmentalism. The three major sections of the book examine conceptual issues, environmental inequities in the United States, and environmental inequities on the international scene.

Shrader-Frechette, Kristin. *Environmental Justice: Creating Equality, Reclaiming Democracy.* **New York: Oxford University Press, 2005.**

The author begins with a somewhat abstruse philosophical analysis of the principles of environmental justice but then goes on to review and analyze a number of problem areas with sharp understanding and perception. She explores land issues in the Appalachians, exposure to chemical toxics in Louisiana, nuclear waste disposal issues in Nevada, and environmental inequities among the Mescalero Apache tribe in the western United States.

Smith, Ted, David A. Sonnenfeld, and David Naguib Pellow, eds. *Challenging the Chip: Labor Rights and Environmental Justice in the Global Electronics Industry.* **Philadelphia: Temple University Press, 2006.**

The production of computer chips, one of the world's most important industries, results in widespread pollution of air, water, and land resources. Those most seriously affected by this pollution tend to be poor, female, minority, and, in many cases, immigrants. This book examines the nature and extent of this environmental inequity in communities around the world.

Stein, Rachel, ed. *New Perspectives on Environmental Justice: Gender, Sexuality, and Activism.* **New Brunswick, NJ: Rutgers University Press, 2004.**

This book examines issues in environmental justice from a feminist and lesbian perspective. It offers both theoretical analyses ("Toward a Queer Ecofeminism" and "Feminist Theory and Environmental Justice") and reviews of practical case studies ("Producing 'Roundup Ready®' communities?" and "No Remedy for the Inuit").

Van Wynsberghe, Robert M. *AlterNatives: Community, Identity, and Environmental Justice on Walpole Island.* Boston: Allyn & Bacon, 2002.

This monograph in the Cultural Survival Studies in Ethnicity and Change series discusses efforts by an Ojibwe community in southern Ontario to protect its unique environmental heritage and the health of its own people from air, water, and land pollution caused by a number of industrial operations in its immediate area. Based on its successes in this effort, the Walpole Island community is placed by the author at the forefront of the environmental justice movement among First Nation people in Canada.

Westra, Laura. *Environmental Justice and the Rights of Indigenous Peoples: International and Domestic Legal Perspectives.* London: Earthscan Publications, 2007.

The author argues that more than 300 million people in over 70 countries worldwide are now being affected by environmental changes that seriously threaten their health and well-being. Moreover, these populations are largely unprotected by laws and regulation, making it possible for corporations to continue environmental inequities with little or no opportunity for legal recourse. She recommends some "modest proposals" for dealing with this situation.

Wilks, Sarah. *Seeking Environmental Justice.* Amsterdam: Rodopi, 2008.

This publication in Rodopi's At the Interface/Probing the Boundaries series consists of articles based on presentations made at the 5th Environmental Justice and Global Citizenship conference held in Oxford, UK, in 2006. Participants represented a wide range of academic disciplines, and their papers covered topics such as environmental education, environmental activism, sustainable communities, genetic modification, free trade and international environmental law, and multinational corporations.

Wulfhorst, J. D., and Anne K. Haugestad, eds. *Building Sustainable Communities: Environmental Justice and Global Citizenship.* Amsterdam: Rodopi, 2006.

The 14 papers that make up this volume were originally presented at the 2004 Ecological Justice and Global Citizenship

Conference in Copenhagen. They discuss issues posed by the development of sustainable communities among populations that are often at risk from environmental inequities. Authors draw on examples from communities in Europe, North America, and Australia.

Articles

The only journal devoted specifically and exclusively to the subject of environmental justice was started in the spring of 2008. *Environmental Justice* is a peer-reviewed quarterly publication of the Mary Ann Liebert, Inc., company. Its ISSN is 1939-4071. Articles in the journal deal with all aspects of the environmental justice movement, such as research on health effects resulting from environmental inequities; efforts to reduce and protect communities from such health effects; analysis of policies and regulations related to issues in environmental justice; the connection between siting of hazardous waste sites and environmental inequities; multidisciplinary studies of the interrelationship of economic, geographical, political, and environmental issues; historical studies of efforts to deal with environmental issues in poor and minority communities; and problems involved in the relocation of hazardous facilities. The journal is intended for a wide variety of readers, including legislators, attorneys, environmentalists, social activists, industry leaders, and public health officials.

Adler, J. H. "There's No Justice in EPA's Environmental Justice Policy." *Corporate Environmental Strategy* **6 (1999): 183–189.**

The author discusses the EPA's Interim Guidance for investigating Title VI of the Civil Rights Act of 1964, issued in 1998, as a mechanism for ensuring that environmental inequities do not occur as a result of the issuance of air quality permits. He finds some serious problems with the regulations.

Agyeman, J., and B. Evans. "'Just Sustainability': The Emerging Discourse of Environmental Justice in Britain?" *The Geographical Journal* **170 (2004): 155–164.**

The authors review the emergence and growth of environmental justice in Great Britain and suggest that the concepts of sustainable development and environmental justice have some common elements (which they call "just sustainability") that can be used to develop public policy and actions for the good of low-income and minority communities as well as the society as a whole.

Baden, B. M., and D. L. Coursey. "The Locality of Waste Sites within the City of Chicago: A Demographic, Social, and Economic Analysis." *Resource and Energy Economics* **24 (2002): 53–93.**

The authors describe this study as a follow-up to the 1987 United Church of Christ study, *Toxic Waste and Race in the United States,* with some of its methodological shortcomings corrected. They find that Hispanics in the city are disproportionately exposed to environmental hazards but that similar patterns for African Americans do not occur, and probably never have existed.

Block, Walter, and Roy Whitehead. "The Unintended Consequences of Environmental Justice." *Forensic Science International* **100 (1999): 57–67.**

Actions taken by the federal government to ensure that environmental inequities do not occur as a result of the siting of industrial facilities may have concurrent harmful economic effects on the communities in question. The authors suggest a number of ways in which industries can make siting decisions that produce less harm to neighboring communities.

Blodgett, Abigail D. "An Analysis of Pollution and Community Advocacy in 'Cancer Alley': Setting an Example for the Environmental Justice Movement in St James Parish, Louisiana." *Local Environment* **11 (2006): 647–661.**

The author investigates the problem of environmental inequities along the 85-mile stretch of the Mississippi River known as Cancer Alley. She finds that the largest concentration of health problems occurs among communities with the highest percentage of African Americans, the lowest average household income, and the lowest rate of high school graduation. She also finds that the best predictors of effectiveness in dealing with environmental justice issues is the ability of a community to network with groups that have expertise in dealing with environmental justice issues, to educate its residents about problems and possible solutions, and to obtain expert legal representation.

Bowen, William M., and Michael V. Wells. "The Politics and Reality of Environmental Justice: A History and Considerations for Public Administrators and Policy Makers." *Public Administration Review* 62 (2002): 688–698.

This article is a brief review of the environmental justice movement intended for public administrators and policy makers, including some recommendations for placing the efforts of the movement on a more rational ground. The authors suggest that issues of environmental inequities can be more effectively dealt with if (1) the quality of empirical research in the field is improved, (2) more emphasis is placed on the difference between "hazard" and "risk" in dealing with environmental inequities, and (3) advocates of environmental justice focus more on public health issues than on "fear, blame, procedural inclusion, and power politics."

Bullard, Robert D., and Glenn S. Johnson. "Environmentalism and Public Policy: Environmental Justice; Grassroots Activism and Its Impact on Public Policy Decision Making." *Journal of Social Issues* 56 (2000): 555–578.

A broad, general introduction to the history of the environmental justice movement. The authors follow the history of the movement from its origins in Warren County, North Carolina, in 1982, with a review of some of its successes and the growth of grassroots organizations designed to deal with environmental inequities.

Cantzler, Julia Miller. "Environmental Justice and Social Power Rhetoric in the Moral Battle over Whaling." *Sociological Inquiry* 77 (2007): 483–512.

The controversy over the Makah Indian tribe's campaign to renew its millennia-old ceremonial whale hunt raises environmental justice questions that are framed from two very different perspectives, which are studied and analyzed in this paper.

Carruthers, David V. "Environmental Justice and the Politics of Energy on the US-Mexico Border." *Environmental Politics* 16 (2007): 394–413.

The environmental justice movement has begun to spread rapidly through Latin America. An example of the issues being raised is the production of a large energy-generation facility in northern Baja California as a way of meeting energy needs in California. The author explores the resistance by environmental justice groups that emerged in Baja to this project.

Charley, Perry, et al. "Navajo Uranium Education Programs: The Search for Environmental Justice." *Applied Environmental Education and Communication: An International Journal* 3 (2004): 101–108.

The authors describe an education program on nuclear materials developed for members of the Navajo tribe in the Four Corners region of the southwestern United States. Since the tribe had no context within which to understand and deal with issues of radioactive materials, new courses had to be developed within the tribal traditions to explain the disproportionate risk to which they were exposed as a result of uranium mining in the area.

Chess, Caron, Joanna Burger, and Melanie McDermott. "Speaking Like a State: Environmental Justice and Fish Consumption Advisories." *Society and Natural Resources* 18 (2005): 267–278.

Government agencies tend to speak in bland, generic language ("government-speak") when issuing warnings about contaminated fish to recreational anglers. As a result, poor and non-English-speaking populations are not aware of the risks posed by such catches and are disproportionately exposed to health hazards caused by eating the fish.

Claudio, Luz. "Standing on Principle: The Global Push for Environmental Justice." *Environmental Health Perspectives* 115 (2007): A500–A503.

The author argues that environmental justice is a growing movement throughout the world. After a brief review of the history of international environmental justice activities, she discusses in some detail current programs for environmental justice in Brazil and South Africa.

Coburn, Jason. "Confronting the Challenges in Reconnecting Urban Planning and Public Health." *American Journal of Public Health* 94 (2004): 541–546.

The author points out that the disciplines of public health and urban planning had a common origin but that they have very little overlap in the modern world. He reviews the history of the two fields and shows how they have grown apart. He then suggests some ways in which they can be reintegrated so as to more effectively deal with problems of environmental inequities.

Cubie, Doreen. "And Clean Water for All." *National Wildlife* **42 (2004): 18–19.**

The author explains the National Wildlife Federation's (NWF) environmental justice efforts to ensure that all Americans have equal access to clean and safe water resources. The article includes a profile of Na'take Osborne, now NWF's national leadership development coordinator.

Derezinski, D. D., M. G. Lacy, and P. B. Stretesky. "Chemical Accidents in the United States, 1990–1996." *Social Science Quarterly* **84 (2003): 122–143.**

The vast majority of research on environmental inequities examines long-term chronic exposure of communities to environmental hazards. This study is one of the very few to ask about the inequities of acute events, such as chemical spills and explosions. The authors find that, although inequities in exposures to acute events do exist, they are less pronounced than in the case of chronic exposures.

Dodds, Lyn, and Bill Hopwood. "BAN Waste, Environmental Justice and Citizen Participation in Policy Setting." *Local Environment* **11 (2006): 269–286.**

In 2000, a group of citizens in Newcastle upon Tyne in England organized to lobby against the construction of a new incinerator in a working-class neighborhood of the city. That group, later given the name of BAN Waste, was eventually successful in preventing construction of the facility. This article explores some of the issues with which BAN Waste dealt and the methods by which they achieved their goal.

Dolinoy, Dana C., and Marie Lynn Miranda. "GIS Modeling of Air Toxics Releases from TRI-Reporting and Non-TRI-Reporting

Facilities: Impacts for Environmental Justice." *Environmental Health Perspectives* 112 (2004): 1717–1724.

Environmental justice researchers usually use records of pollution produced by industries that release large quantities of hazardous materials into the atmosphere (TRI reporters). The authors decided to also use data from smaller polluters, so-called non-TRI reporters. They found that environmental inequities as measured by data from non-TRI reporters were often much greater than those calculated with TRI data. They conclude that these results suggest some important changes in the way environmental justice research should be conducted.

Downey, Liam, and Marieke Van Willigen. "Environmental Stressors: The Mental Health Impacts of Living Near Industrial Activity." *Journal of Health and Social Behavior* 3 (2005): 289–305.

Although the health effects of environmental inequities have been rather widely studied, little research has been done on the mental health effects of poor and ethnic populations living near hazardous sites. In this study, Downey and Van Willigen found that such populations are significantly more likely to suffer mental health problems, much as they do physical health problems.

Edwards, B., and A. E. Ladd. "Environmental Justice, Swine Production and Farm Loss in North Carolina." *Sociological Spectrum* 20 (2000): 263–290.

North Carolina has had the largest growth of swine farms over the last decade in the United States. This growth has had deleterious effects on land, air, and water and has led to the loss of large amounts of farmland. The authors report that African American and low-income communities have disproportionately been affected by these trends in the state.

Elliott, M. R., et al. "Environmental Justice: Frequency and Severity of US Chemical Industry Accidents and the Socioeconomic Status of Surrounding Communities." *Journal of Epidemiology and Community Health* 58 (2004): 24–30.

The authors study the relationship between certain types of industrial accidents and demographic characteristics of the communities adjacent to the industrial facilities. As one might predict, they find that chemical and industrial plants are sited near minority communities more often than near white communities. They also discover, however, that industrial facilities located near minority communities are more likely to have accidents than are those near white communities. They recommend that greater efforts be made not only to more fairly distribute the location of hazardous industrial facilities but also to better enforce regulations designed to reduce the likelihood of accidents.

Elliott, S. J., et al. "The Power of Perception: Health Risk Attributed to Air Pollution in an Urban Industrial Neighbourhood." *Risk Analysis* 19 (1999): 621–634.

The authors report on environmental justice action in a Hamilton, Ontario, Canada, neighborhood using health assessment results from a group of experts in the field, on the one hand, and from a community survey, on the other. They point out that residents had a greater degree of confidence in the results obtained by the latter method.

Faber, D., and D. McCarthy. "The Evolving Structure of the Environmental Justice Movement in the United States: New Models for Democratic Decision-Making." *Social Justice Research* 14 (2001): 405–421.

The authors point out that mainstream environmental organizations in the United States have, over the last few decades, reduced rather than expanded the participation of ordinary citizens in the environmental movement. By contrast, the environmental justice movement has moved in just the opposite direction, bringing more and more diverse individuals into the movement. The authors claim that this pattern in the long run will ultimately result in more tolerable levels of environmental hazards for more affluent populations and continued environmental battles for low-income and minority communities. They recommend new ways of conceptualizing environmental issues to avoid this scenario.

Fan, Mei-Fang. "Nuclear Waste Facilities on Tribal Land: The Yami's Struggles for Environmental Justice." *Local Environment* 11 (2006): 433–444.

The striking similarities of struggles for environmental justice in many parts of the world are illustrated by this case study of the battle by members of the Yami tribe against the construction of a nuclear waste depository on their land on Orchid Island, Taiwan.

Fricker, R. D., and N. W. Hengartner. "Environmental Equity and the Distribution of Toxic Release Inventory and Other Environmentally Undesirable Sites in Metropolitan New York City." *Environmental and Ecological Statistics* **8 (2001): 33–52.**

The authors study 354 "environmentally undesirable facilities" located in or adjacent to 2,216 census tracts in New York City to determine if there is a disproportionate exposure of ethnic groups to environmental hazards. They find differing patterns in each of the city's boroughs with, in general, Hispanic neighborhoods more likely to be associated with hazard sites than African American neighborhoods, and both groups at greater risk than white neighborhoods.

Gandy, M. "Between Borinquen and the Barrio: Environmental Justice and New York City's Puerto Rican Community, 1969–1972." *Antipode* **34 (2002): 730–761.**

Although most historians trace the origins of the environmental justice movement in the United States to the early 1980s, precursors of the movement appeared much earlier in history. This book describes the work of a radical Puerto Rican organization called the Young Lords, which fought for environmental justice in a number of cities in the late 1960s and early 1970s.

Godsil, Rachel D. "Remedying Environmental Racism." *Michigan Law Review* **90 (1991): 394–427.**

This article has special significance because it appears to be the first note written in a law journal about the topic of environmental justice. The author provides a detailed review of the history of the environmental justice movement, from its beginnings in Warren County, North Carolina, in 1982, along with an analysis of legal issues involved in the identification, analysis, and remediation of problems related to environmental inequities.

Gouldson, Andy. "Do Firms Adopt Lower Standards in Poorer Areas? Corporate Social Responsibility and Environmental Justice in the EU and the US." *Area* **38 (2006): 402–412.**

Using new data on pollution produced by oil refineries in the United States and the European Union, the author asks (1) how international environmental standards adopted by corporations are translated on a local scale, (2) whether implementation of these standards differs from place to place, and (3) whether the implementation of standards reflects any understanding and application of environmental justice principles.

Gowda, M. V. Rajeev, and Doug Easterling. "Voluntary Siting and Equity: The MRS Facility Experience in Native America." *Risk Analysis* **20 (2000): 917–930.**

The U.S. government's monitored, retrievable storage (MRS) program was a plan for the siting of nuclear waste disposal sites in various parts of the western United States. The program eventually became focused on the use of Native American lands for the siting of nuclear dumps, resulting in a complex and diverse set of problems for various tribes. The authors review the history of the MRS program and analyze its implications for environmental justice among Native American populations.

Griffith, Matt, Mansoureh Tajik, and Steve Wing. "Patterns of Agricultural Pesticide Use in Relation to Socioeconomic Characteristics of the Population in the Rural U.S. South." *International Journal of Health Services* **37 (2007): 259–277.**

The authors attempted to assess the amount spent by agricultural companies on pesticides in a county compared to the racial composition and income of residents in the county. They found very significant relationships, with up to eight times as much spent on pesticides for low-income communities with a high percentage of nonwhite residents.

Grineski, Sara, Bob Bolin, and Christopher Boone. "Criteria Air Pollution and Marginalized Populations: Environmental Inequity in Metropolitan Phoenix, Arizona." *Social Science Quarterly* **88 (2007): 535–554.**

The authors suggest that new research technologies are available for assessing the effects of air pollutants on specific human populations. They employ these technologies to assess the differential effects of air pollution on residents of Phoenix. They find that neighborhoods with a large proportion of Latino immigrants, low-income individuals, and renters (but not African Americans) tend to have higher rates of health problems associated with air pollutants.

Hooks, Gregory, and Chad L. Smith. "The Treadmill of Destruction: National Sacrifice Areas and Native Americans." *American Sociological Review* **69 (2004): 558–575.**

The authors argue that in the history of the environmental justice movement the special problems of Native Americans have been largely ignored. They point to the widespread use of Native American lands for waste dumps by the U.S. military and provide data to show how even abandoned bases pose serious and disproportionate threats to Indian communities.

Hornberg, Claudia, and Andrea Pauli. "Child Poverty and Environmental Justice." *International Journal of Hygiene and Environmental Health* **210 (2007): 571–580.**

The authors consider the special question of the effects of environmental inequities on the health of children and suggest that greater attention be paid to this issue within the environmental justice movement and by governmental agencies.

Illsley, B. M. "Good Neighbour Agreements: The First Step to Environmental Justice?" *Local Environment* **7 (2002): 69–79.**

Good neighbor agreements are contracts drawn up between industries and the communities in which they are located promising to promote environmental justice practices in the community. The author explores the good neighbor agreement drawn up between Dundee Energy Recycling Limited and the city of Dundee in the United Kingdom. She suggests that the agreement appears to hold the potential for preventing environmental inequities but that the long-term effectiveness of the tool has not yet been demonstrated.

Ishiyama, N. "Environmental Justice and American Indian Tribal Sovereignty: Case Study of a Land-Use Conflict in Skull Valley, Utah." *Antipode* **35 (2003): 119–139.**

The author describes a case in which the Skull Valley Band of Goshute Indians agrees to host a nuclear waste disposal site on its lands and explains why the case is far more complex than it appears to be at first glance. Native Americans have, the author argues, fewer options than are at first apparent, making their choices on issues of environmental inequity more difficult.

Koenig, Thomas H., and Michael L. Rustad. "Toxic Torts, Politics, and Environmental Justice: The Case for Crimtorts." *Law & Policy* 26 (2004): 189–207.

The authors point out that so-called crimtorts, which include penalties calibrated to the wealth of a guilty party, are currently under attack by neoconservatives. The loss of such penalties, they say, may greatly reduce the effectiveness of many actions taken by environmental justice groups.

Krieg, Eric. "Race and Environmental Justice in Buffalo, NY: A ZIP Code and Historical Analysis of Ecological Hazards." *Society and Natural Resources* 18 (2005): 199–213.

An analysis of hazardous waste disposal patterns reveals evidence that low-income populations are disproportionately affected, whereas such patterns are not observed for various ethnic groups. The author explains that these differences are a result of historic patterns of housing discrimination in the city.

Krieg, E. J., and D. R. Faber. "Not So Black and White: Environmental Justice and Cumulative Impact Assessments." *Environmental Impact Assessment Review* 24 (2004): 667–694.

While acknowledging that the specific impacts of environmental inequities on minority and low-income communities have been well demonstrated, the authors suggest that a more comprehensive analysis of such impacts would be useful. In using this type of analysis of a Massachusetts community, they found a linear relationship between minority composition of a community and environmental inequities over a broad range of compositions.

Kushmerick, Ann, Lindsay Young, and Susan Stein. "Environmental Justice Content in Mainstream US, 6–12 Environmental Education Guides." *Environmental Education Research* 13 (2007): 385–408.

The authors ask to what extent the principles of environmental justice are incorporated into national programs of environmental education, such as Project WILD, Project Learning Tree, and

Project WET. They find that, while the programs often address issues of importance in environmental justice, they seldom make explicit connections between those principles and the environmental justice movement itself.

Lavelle, Marianne, and Marcia Coyle. "Unequal Protection: The Racial Divide in Environmental Law." *National Law Journal* **21 (1992): 81–92.**

A very important historical document in which the authors report on an eight-month-long investigation of actions and decisions taken by the EPA on issues of environmental inequities in low-income and minority communities. Researchers concluded that the EPA did not provide equal treatment to polluters and that there is a "racial divide" in EPA's policies and actions in dealing with environmental problems.

Maantay, Juliana. "Asthma and Air Pollution in the Bronx: Methodological and Data Considerations in Using GIS for Environmental Justice and Health Research." *Health & Place* **13 (2007): 32–56.**

The author considers some methodological issues in using the Geographic Information Systems (GIS) program for the study of environmental problems and reports on the use of GIS for the study of asthma and air pollution in the Bronx, New York City. She finds that people living near noxious land-use sites are 66 percent more likely to be hospitalized for asthma that those outside the area.

Maclachlan, John C., et al. "Mapping Health on the Internet: A New Tool for Environmental Justice and Public Health Research." *Health & Place* **13 (2007): 72–86.**

The authors describe the use of Internet-based Geographic Information Services (Web-GIS) for collecting data needed for studies on environmental inequities, thus alleviating problems of cost and data availability that often arise in such studies.

McCarthy, L. "The Brownfield Dual Land-Use Policy Challenge: Reducing Barriers to Private Redevelopment While Connecting Reuse to Broader Community Goals." *Land Use Policy* **19 (2002): 287–296.**

The development of brownfields for beneficial purposes raises a number of social, economic, and environmental questions. While the recovery of previously unusable land is largely beneficial to the general community, the potential health risks of the brownfields pose concerns for people living in the area. Policies must be developed that allow development without creating risks for the community as a whole or for specific nearby communities.

Mohai, Paul, and Robin Saha. "Reassessing Racial and Socioeconomic Disparities in Environmental Justice Research." *Demography* **43 (2006): 383–399.**

Studies on environmental inequities tend to show that low-income and minority populations bear an unequal share of environmental hazards. But those studies show a widely variable correlation between health risk and race or income. The authors suggest some improvements in research methodology that can result in clearer relationships among the variables being studied.

Morello-Frosch, Rachel, et al. "Environmental Justice and Regional Inequality in Southern California: Implications for Future Research." *Environmental Health Perspectives* **110 (April 2002): 149–154.**

The authors discuss some of the methodological issues involved in assessing the quantitative relationships between pollution sources and environmental inequities, and they report on a study on the health effects of air pollution in Southern California. They conclude that communities of color "bear a disproportionate burden in the location of treatment, storage, and disposal facilities and Toxic Release Inventory facilities."

Murphy-Greene, Celeste, and Leslie Leip. "Assessing the Effectiveness of Executive Order 12898: Environmental Justice for All?" *Public Administration Review* **62 (2002): 679–687.**

The authors ask how well laws and regulations designed to protect farmworkers in Florida from exposure to pesticides are working. They conclude that the laws and regulations are not implemented very effectively, and farmworkers have little understanding of the protections the laws and regulations are supposed to provide.

Newell, Peter. "Trade and Environmental Justice in Latin America." *New Political Economy* **12 (2007): 237–259.**

The author explores the relationship between changing trade policies that involve the expansion of free trade among nations and environmental justice issues. He points out that environmental justice issues are increasingly a part of agreements made between and among countries. This trend, he believes, creates both potential benefits and risks for environmental justice groups in less-developed nations.

Norton, Jennifer M., et al. "Race, Wealth, and Solid Waste Facilities in North Carolina." *Environmental Health Perspectives* **115 (2007): 1344–1350.**

This report on the distribution of solid waste disposal sites in North Carolina finds that low-income communities and those with large percentages of people of color are much more likely to be situated near waste sites.

Nussbaum, Rudi, et al. "Community-Based Participatory Health Survey of Hanford, WA, Downwinders: A Model for Citizen Empowerment." *Society and Natural Resources* **17 (2004): 547–559.**

As early as the mid-1940s, people living near the Hanford, Washington, nuclear research facility began to experience health problems that, they believed, were associated with the facility. This article reviews the long battle by the Hanford Downwinders to obtain evidence for this connection and to force the federal government to take action to deal with the problem. The authors suggest that the success of the Downwinders may serve as a model for other groups working for environmental justice in their own communities.

O'Neil, Sandra George. "Superfund: Evaluating the Impact of Executive Order 12898." *Environmental Health Perspectives* **115 (2007): 1087–1093.**

The author reviews the status of 1,540 sites selected for cleanup under the Superfund program of the EPA before and after Executive Order 12898 in 1994 and finds that the order has had no discernible effect in increasing the rate at which environmental inequities are being addressed.

Pastor, Manuel, Rachel Morello-Frosch, and James L. Sadd. "Breathless: Schools, Air Toxics, and Environmental Justice in California." *Policy Studies Journal* 34 (2006): 337–362.

This study confirms that air pollution has a disparate effect on the academic achievements of children of color in California. The authors suggest the need for special remediation in "hot spots" where pollution is most serious.

Pastor, Manuel, Jim Sadd, and John Hipp. "Which Came First? Toxic Facilities, Minority Move-In, and Environmental Justice." *Journal of Urban Affairs* 23 (2001): 1–21.

This article addresses one of the fundamental issues in the study of environmental inequities: Do people move into areas that are already hazardous, or do industries choose to place hazardous facilities in existing minority and low-income communities? The authors perform a sophisticated analysis on a community in Los Angeles on this question and conclude that the latter appears to be the case in this instance.

Pellow, D. N., A. Weinberg, and A. Schnaiberg. "The Environmental Justice Movement: Equitable Allocation of the Costs and Benefits of Environmental Management Outcomes." *Social Justice Research* 14 (2001): 423–439.

The authors suggest a number of changes in the nature of environmental justice research, including greater attention to the historical context of disputes, a better understanding of the role of all stakeholders in a controversy, possible trade-offs between environmental protection and social equity, and the impact of social movement activity on the state of environmental protection.

Petersen, Dana, et al. "Community-Based Participatory Research as a Tool for Policy Change: A Case Study of the Southern California Environmental Justice Collaborative." *Review of Policy Research* 23 (2006): 339–354.

This article explains how research by the Southern California Environmental Justice Collaborative, an environmental justice group, was instrumental in the state of California's reduction of its allowable risk levels for stationary sources by 75 percent. It discusses the implications of this experience for other environmental justice groups.

Petrie, Michelle. "Environmental Justice in the South: An Analysis of the Determinants and Consequences of Community Involvement in Superfund." *Sociological Spectrum* **26 (2006): 471–489.**

The study reported here examines demographic factors involved in communities where Superfund cleanups have taken place. Research shows that cleanups occurred more slowly and were more likely to have deleterious effects in communities with the greatest level of environmental justice activism, and that communities with higher percentages of minority groups were less likely to become involved in cleanup efforts.

Petts, Judith. "Enhancing Environmental Equity through Decision-Making: Learning from Waste Management." *Local Environment* **10 (2005): 397–409.**

A study conducted in Los Angeles finds that access to park space is significantly less for low-income, African American, Pacific Islander, and Latino populations than for predominantly white populations. The author points out some practical problems involved in resolving this environmental inequity.

Postma, Julie. "Environmental Justice: Implications for Occupational Health Nurses." *AAOHN Journal* **54 (2006): 497–498.**

The author suggests that occupational health nurses have special opportunities to discover, assess, and deal with environmental justice issues in their daily work. She outlines some methods that nurses can use in integrating environmental justice principles into their occupation, including the mapping of hazardous risks, developing strategic plans for dealing with risks, and making available their specialized skills in working with grassroots organizations.

Powell, Dana E. "Technologies of Existence: The Indigenous Environmental Justice Movement." *Development* **49 (2006): 125–132.**

The author suggests that new emphases on wind and solar energy present indigenous peoples with an alternative to the exploitation of their lands by the petroleum industry and serve as a way of handling the environmental inequities with which the tribes have traditionally had to deal.

Ranco, Darren, and Dean Suagee. "Tribal Sovereignty and the Problem of Difference in Environmental Regulation: Observations on 'Measured Separatism' in Indian Country." *Antipode* 39 (2007): 691–707.

Since American Indian tribes are supposed to have legal and judicial sovereignty, they apparently have the right to take actions necessary to reduce or eliminate environmental inequities that affect their communities. U.S. courts, however, have often attempted to abridge those rights. The authors argue for a stronger position on American Indian sovereignty and suggest that principles involved in making this claim may be applicable to other minority groups facing issues of environmental injustice.

Rogge, Mary E., et al. "Leveraging Environmental, Social, and Economic Justice at Chattanooga Creek: A Case Study." *Journal of Community Practice* 13 (2005): 33–53.

This case study describes the successful efforts by an African American community located on the most polluted river in the southeastern United States to reduce the contamination in the waterway. The authors analyze the sociological and other factors involved in the group's efforts.

Schlosberg, David. "Reconceiving Environmental Justice: Global Movements and Political Theories." *Environmental Politics* 13 (2004): 517–540.

The author argues that little effort has been expended in explaining what people mean by "justice" when they refer to environmental justice. The term should, he says, refer to more than equal distribution of environmental hazards. A clarification of the meaning of the term, he goes on, has significant implications for the extension of environmental justice throughout the world.

Shulman, Stuart W., et al. "Empowering Environmentally-Burdened Communities in the US: A Primer on the Emerging Role for Information Technology." *Local Environment* 10 (2005): 501–512.

The authors discuss the use of information technology (IT) in dealing with problems of environmental injustice. They argue that disproportionate knowledge about and access to IT is an important issue for activists dealing with such issues.

Sobotta, Robin R., Heather E. Campbell, and Beverly J. Owens. "Aviation Noise and Environmental Justice: The Barrio Barrier." *Journal of Regional Science* 47 (2007): 125–154.

The authors suggest that exposure to aviation noise is a possible manifestation of environmental inequities, and they conduct a study on this topic. They find, as in other forms of environmental inequities, that minority communities are more likely to suffer health effects from living near sources of noise.

Sperber, Irwin. "Alienation in the Environmental Movement: Regressive Tendencies in the Struggle for Environmental Justice." *Capitalism, Nature, Socialism* 14 (2003): 1–43.

This article suggests the very interesting hypothesis that environmental justice organizations are increasingly being co-opted by capitalist forces because of their increasing dependence on corporate foundation grants and their interaction with corporate foes and government regulators. He suggests that organizations need to restore their traditional connection with and dependence on grassroots organizations and forces.

Tajik, Mansoureh, and Meredith Minkler. "Environmental Justice Research and Action: A Case Study in Political Economy and Community-Academic Collaboration. *International Quarterly of Community Health Education* 26 (2006–2007): 213–231.

Partnerships between academic researchers and grassroots community activists are increasingly common in efforts to reduce environmental inequities in minority and low-income communities. This paper reviews such a collaboration between Concerned Citizens of Tillery (North Carolina) and the University of North Carolina, Chapel Hill, School of Public Health to deal with the health problems posed by the siting of a large-scale hog operation in Halifax County.

Toffolon-Weiss, Melissa, and J. Timmons Roberts. "Toxic Torts, Public Interest Law, and Environmental Justice: Evidence from Louisiana." *Law & Policy* 26 (2004): 259–287.

Some poor and minority communities that take action against industries that pollute their neighborhoods call on assistance from private injury lawyers, while others go to public interest groups.

The authors ask whether this choice makes any difference in the ultimate results obtained by the groups.

Warner, K. "Linking Local Sustainability Initiatives with Environmental Justice." *Local Environment* **7 (2002): 35–47.**

Over the last decade, a number of people have attempted to link the goals and actions of the environmental justice movement with the movement for sustainable development. In this paper, the author reports on a survey of 33 large cities in the United States to see which and how many consciously incorporate the principles of environmental justice into their programs for the development of sustainable communities in the city.

Williams, Robert W. "Environmental Injustice in America and Its Politics of Scale." *Political Geography* **18 (1999): 49–73.**

The author argues that laws and regulations in the United States treat environmental injustice as if it were a national, or at least widespread, problem. He claims, however, that nearly all instances of environmental inequity are essentially local, making laws and regulations irrelevant or ineffective. He concludes that government regulation may be largely unnecessary in dealing with cases of environmental inequities.

Wolch, Jennifer, John P. Wilson, and Jed Fehrenbach. "Parks and Park Funding in Los Angeles: An Equity-Mapping Analysis." *Urban Geography* **26 (2005): 4–35.**

The authors not only discover significant environmental inequities in Los Angeles public park siting but also find endemic factors in city funding policies that perpetuate these patterns.

Zaferatos, Nicholas. "Environmental Justice in Indian Country: Dumpsite Remediation on the Swinomish Indian Reservation." *Environmental Management* **38 (2006): 896–909.**

Members of the Swinomish tribe struggled for more than two decades to convince the EPA to require PM Northwest, Inc., to clean up a hazardous waste dump on its reservation. The author uses this case study to review some fundamental problems of environmental justice with which Indian communities have had to deal.

Reports

American Lung Association. *State of Lung Disease in Diverse Communities: 2007.* New York: American Lung Association, 2007.

The American Lung Association has conducted an exhaustive survey on respiratory problems in a variety of specialized communities, including those of African Americans, Native Americans, Puerto Ricans, Native Hawaiians and other Pacific Islanders, Asian Americans, and Hispanic Americans. The organization's research confirms that such health problems tend to occur significantly more often among minority groups than among white Americans.

Bonorris, Steven. *Environmental Justice for All: A Fifty-State Survey of Legislation, Policies, and Initiatives.* San Francisco: Hastings College of the Law, University of California, January 2004.

This report is the third edition of the most complete and comprehensive survey of legislation, policies, and initiatives in the 50 states. For each state, it provides a summary of statutes, policies, partnership agreements, and other documents, along with a list of contacts for queries about environmental justice issues.

Bullard, Robert D., Paul Mohai, Robin Saha, and Beverly Wright. *Toxic Wastes and Race at Twenty: 1987–2007.* Cleveland, OH: United Church of Christ, March 2007.

This report is a follow-up to the classic 1987 study, *Toxic Wastes and Race in the United States,* which documented the widespread existence of environmental inequities within low-income and minority communities. The purpose of this study was to determine the progress made in the 20 years since the original study was published. The researchers conclude that conditions in 2007 are "very much the same as they were in 1987."

Committee on Environmental Justice, Institute of Medicine. *Toward Environmental Justice: Research, Education, and Health Policy Needs.* Washington, DC: National Academies Press, 1999.

One of the classic studies in environmental justice, this report reviews five case studies in environmental inequities, identifies hazards facing low-income and minority communities, suggests methods for studying problems of environmental inequity, discusses some essential characteristics of environmental education, and recommends a number of actions to be taken by decision makers and policy makers.

Office of Inspector General, Environmental Protection Agency. *EPA Needs to Conduct Environmental Justice Reviews of Its Programs, Policies, and Activities.* **Washington, DC: Environmental Protection Agency, September 18, 2006.**

The inspector general's review team finds that senior managers at the EPA have not sufficiently directed subordinate offices to conduct environmental justice reviews and that, as a result, the majority of those offices have not as yet conducted such reviews.

Office of Inspector General, Environmental Protection Agency. *EPA Needs to Consistently Implement the Intent of the Executive Order on Environmental Justice.* **Washington, DC: Environmental Protection Agency, March 1, 2004.**

Ten years after President Bill Clinton issued Executive Order 12898, the EPA's Office of the Inspector General conducts a study to see how effectively the agency is carrying out the requirements of the order. It finds that the agency has made little progress in achieving the objectives of the order and, in fact, has reinterpreted the meaning of the order to *exclude* minorities and low-income populations.

U.S. Commission on Civil Rights. *Not in My Backyard: Executive Order 12898 and Title VI as Tools for Achieving Environmental Justice.* **Washington, DC: U.S. Commission on Civil Rights, October 2003.**

This report examines how well four federal agencies—the EPA, the U.S. Department of the Interior, the U.S. Department of Housing and Urban Development, and the U.S. Department of Transportation—have implemented Executive Order 12898 and Title VI of the 1964 Civil Rights Act. The authors of the report conclude that the answer to that question is: "Not very well."

U.S. Government Accountability Office. *Environmental Justice: EPA Should Devote More Attention to Environmental Justice When Developing Clean Air Rules.* **Washington, DC: Government Accountability Office, July 2005.**

At the request of the ranking member of the Subcommittee on Environment and Hazardous Materials of the House Committee on Energy and Commerce, the Government Accountability Office (GAO) undertook a study to determine whether the EPA had taken into consideration the principles of environmental justice in drafting three new regulations on air quality control. The GAO found that the EPA, in contradiction to the provisions of Executive Order 12898, had essentially ignored concerns about environmental inequities in drafting those regulations.

Internet Sources

American Bar Association, Section on Energy, Environment, and Resources. "The Law of Environmental Justice: Update Service." Available at http://www.abanet.org/environ/committees/envtab/ejweb.html.

This Web site may be the best single source of information on laws, regulations, court cases, and other legal issues related to environmental justice. Citations date as far back as 1998.

Barbalance, Roberta C. "Environmental Justice and the NIMBY Principle." Available at http://environmentalchemistry.com/yogi/hazmat/articles/nimby.html.

The author discusses the issue of environmental inequities within the larger context of the well-known principle of "not in my backyard" (NIMBY).

Collado, Jonathan, et al. "Environmental Justice through Youth Empowerment." Available at http://www.ejconference2008.org/images/Chelsea_Creek.pdf.

Ten members of an environmental justice movement in metropolitan Boston, the East Boston Environmental Chelsea Creek Crew (all under the age of 30), describe their environmental justice activities and explain how their concept of environmental justice differs from those who have been involved in the movement over much longer periods of time.

Ecofeminism.net. "Environmental Racism." Available at http:// www.ecofeminism.net/content/environmental_racism.htm.

This Web site provides links to a number of publications and Internet sites with information about environmental racism and environmental justice topics.

EJCONFERENCE2008.ORG. Available at http://www.ejconference2008.org/.

EJCONFERENCE2008.ORG is the official Web site of the 2008 Environmental Justice in America Conference held at Howard University, Washington, D.C., on May 21–24, 2008. Presentations at the conference discussed success stories in the environmental justice movement; corporate/community relations; environmental justice implications in the Hurricane Katrina response; just and sustainable energy policies; community participation in environmental decision making; environmental health in disadvantaged communities; environmental education; and innovations in environmental justice legislation, regulation, and litigation.

Endres, Danielle. "The State of Environmental Justice in High-Level Nuclear Waste Siting Decisions." Available at http:// www.ejconference2008.org/images/Endres.pdf.

This paper was presented at the 2007 Environmental Justice Conference at Howard University. The author reviews events in the siting of nuclear waste disposal sites on Indian lands and suggests some further steps that can be taken to reduce environmental inequities in such actions.

Environmental Education and Training Partnership. "Environmental Justice and Environmental Education." Available at http://eelink.net/eetap/info59.pdf.

This short paper explains what environmental justice is and discusses some issues on including environmental justice in programs of environmental education.

"Environmental Justice/Environmental Racism." Available at http://www.ejnet.org/ej/.

This Web site provides a general introduction to environmental justice, lists some important documents in the field, and provides links to other Web sites and resources and to information on government policies on environmental justice.

"The Environmental Justice Information Page." Available at http://eelink.net/EJ/.

This Web site was prepared by members of the Ecological Issues class at the University of Michigan's School of Natural Resources and Environment. Although now outdated, it still provides some useful information on basic aspects of the environmental justice movement.

Environmental Justice Resource Center at Clark Atlanta University. Available at http://www.ejrc.cau.edu/.

This Web site is a treasure trove of information on the environmental justice movement, with lists of books, articles, reports, curriculum guides, and other documents on environmental justice; articles on a variety of specialized topics related to environmental inequities; news and events related to environmental justice; and links to other sites with information on environmental justice.

Federal Highway Administration and Federal Transit Administration. "Environmental Justice." Available at http://www .fhwa.dot.gov/environment/ej2.htm.

Of all federal agencies, the Federal Highway Administration and Federal Transit Administration appear to have been among the most active in developing programs of environmental justice for which they are responsible. Their Web site provides an excellent review of those programs as well as a good general introduction to the field of environmental justice as it relates to transportation issues.

"Indian Land and Dumpsites." Available at http://nativenet .uthscsa.edu/archive/nl/91c/0053.html.

This now outdated blog provides an excellent introduction to the problems faced by Indian tribes that are offered the opportunity to host hazardous waste sites on their lands. The episode described here is only one of many similar events that occurred on Indian reservations during the 1990s.

Institute of Transportation Studies, University of California at Berkeley. "Environmental Justice & Transportation: A Citizen's Handbook." Available at http://www.its.berkeley.edu/research/ ejhandbook/ejhandbook.html.

This Web site provides access to the institute's booklet. The booklet provides a general introduction to environmental justice and discusses its application to problems of transportation policy and planning.

Jantz, Eric. "Environmental Racism with a Faint Green Glow: The U.S. Nuclear Regulatory Commission's Missed Opportunity to Create a Meaningful Environmental Justice Policy." Available at http://www.ejconference2008.org/images/Jantz.pdf.

The author reviews the development of an environmental justice policy by the U.S. Nuclear Regulatory Commission and discusses shortcomings in that policy.

MapCruzin.com. "Environmental Justice." Available at http:// www.mapcruzin.com/environmental_justice.htm.

This Web site focuses on mapping, education, and research related to Geographic Information System (GIS) technology. The environmental justice page has links to organizations, issues, academics, bibliographies, articles, agencies, and research in the field that use GIS.

Nauman, Talli. "Indigenous Environmental Justice Issues Enter the Global Ring." Available at http://americas.irc-on-line.org/am/837.

The author presents a superb introduction to the rise of environmental justice among indigenous people, both in the United States and other parts of the world; discusses some issues that are of particular interest to such communities; and provides an extensive set of references and links to U.S. and international governmental and nongovernmental groups working in the field of environmental justice for indigenous communities.

New York State, Department of Environmental Conservation, Office of Environmental Justice. "Environmental Justice." Available at http://www.dec.ny.gov/public/333.html.

Many states now have offices, departments, divisions, or other units with specific responsibility for environmental justice issues in the state. The New York State office is one of the largest, best organized, and most user friendly. This Web site describes the office's organization and functions in detail.

Robinson, Deborah M. "Environmental Racism: Old Wine in a New Bottle." Available at http://www.wcc-coe.org/wcc/what/ jpc/echoes/echoes-17-02.html.

This article reviews the history of the environmental justice movement in the United States, provides examples of environmental racism in both the United States and the rest of the world, and discusses the World Conference against Racism held in Durban, South Africa, on August 31 to September 7, 2001.

Scorecard: The Pollution Information Site. "Environmental Justice." Available at http://www.scorecard.org/community/ej-index.tcl.

This Web site is an outstanding source of information about environmental inequities for every part of the United States. For any zip code area, it provides data on the distribution of environmental burdens by race, ethnicity, income, occupation, and other variables; comparisons for these variables for other parts of the same state; and detailed information on related topics, such as Superfund sites, toxic chemicals, and air pollutants.

Sierra Club. "Environmental Justice." Available at http://www .sierraclub.org/ej/.

This section of the Sierra Club Web site provides information on the organization's environmental justice regional projects; its newsletter, *EJ Activist;* and news about the environmental justice movement and upcoming events.

Stokes, Lance, and Kenneth L. Green. "Twenty-Five Years of 'Change' and Things Remain the Same." Available at http:// www.ejconference2008.org/images/Green_Stokes.pdf.

The authors point out that remediation of brownfield sites is required by law to include black-owned firms but that government agencies over the past few decades have consistently ignored that requirement.

Sze, Julie. "Toxic Soup Redux: Why Environmental Racism and Environmental Justice Matter after Katrina." Available at http://understandingkatrina.ssrc.org/Sze/.

The author points out that most discussions of Hurricane Katrina and cleanup efforts following that disaster have largely ignored issues of environmental racism, which have strong bearing on the failures of efforts to help New Orleans residents recover from the event.

University of Michigan, School of Natural Resources and Environment. "Environmental Justice Initiative." Available at http://eji.snre.umich.edu/.

The University of Michigan's School of Natural Resources and Environment has long been an important center of research and activism in the field of environmental justice. This Web page describes the school's Environmental Justice Initiative, courses offered in the school, research related to environmental inequities, and events related to environmental justice.

U.S. Environmental Protection Agency. "Environmental Justice." Available at http://www.epa.gov/compliance/environmen taljustice/.

This page is the starting point for all EPA documents dealing with environmental justice, including information on the National Environmental Justice Advisory Committee, the Federal Interagency Working Group, and grants available in the field of environmental justice.

Wikipedia. "Environmental Justice." Available at http://en.wiki pedia.org/wiki/Environmental_justice.

This Wikipedia entry provides a good general introduction to the topic of environmental justice, with a number of recommended references with which to follow up.

Glossary

acceptable risk   Possible hazard whose magnitude is considered to be small enough to accept in exchange for some valuable outcome.

ambient   An adjective referring to surrounding conditions. Ambient air, for example, is the surrounding air.

BANANA (building absolutely nothing anywhere near anything) principle   A term ascribed to some environmentalists who would oppose the construction of any environmentally harmful facility anywhere because of its potential effects on plant, animal, and human life. *See also* **NIMBY** and **NOPE**.

benign neglect   A policy of ignoring an unpleasant situation rather than finding ways to deal with the situation.

best available technology (BAT)   The most efficient and/or effective way of dealing with some environmental issue, such as the control of air pollution, no matter what the cost of that technology may be.

blaming the victim   A philosophy that bad things happen to people because it is their own fault.

boomerang effect   An effect that may occur when pesticides or other chemicals banned in the United States are shipped to other nations and then returned to this country, causing unanticipated health effects.

boycott   A nonviolent form of protest in which a group of people jointly agree not to purchase a product or patronize an establishment to demonstrate their opposition to some policy or action supported by the manufacturer or establishment.

brownfield   An abandoned industrial site where soil and groundwater have been polluted.

burden of proof   In law, the determination as to who must prove that harm has or has not, will or will not, be caused by some practice.

business as usual The tendency of a company or the government to continue conducting business as it has in the past without regard to conditions that might reasonably warrant changes in that business pattern.

buyout To completely purchase a home, business, or other operation.

capability analysis A study of the ways in which a piece of land can be used.

circle of poison *See* **boomerang effect**.

civil disobedience The refusal to obey certain laws and regulations to protest governmental or other policies.

class action suit A legal action taken on behalf of all members of a group directly affected by a case.

clawback agreement A provision by which one member of an agreement is able to collect compensation, usually in the form of money, from a second party to the agreement if and when that second party does not fulfill its obligations according to the terms of that agreement.

cleanup The actions taken to neutralize and/or remove hazardous wastes from an area.

community buyout The practice by which a governmental body or private industry pays for the complete removal of a community because of its proximity to one or more environmentally dangerous facilities.

correlation A statistical term that refers to the extent to which two variables are mathematically related to each other. Correlation does not necessarily prove causation.

cost/benefit analysis Any attempt to compare the advantages of taking some action (such as installing a waste disposal incinerator) with the disadvantages of that action (such as the release of harmful gases from the incinerator).

cost effectiveness analysis An attempt to compare the effectiveness and desirability of achieving an improvement in environmental conditions (such as a reduction in air pollution) by various technologies.

DAD (decide, announce, defend) An acronym used to describe a common industrial practice of making decisions about the siting of environmentally harmful facilities.

de facto Latin for "by fact," referring to practices and policies that actually exist, whether or not they are established and maintained by laws and regulations.

de jure Latin for "by right," referring to practices and policies that are established and maintained by law and regulation.

debt-for-nature swap A mechanism by which poor countries agree to set aside parts of their land for conservation purposes in exchange for the cancellation of all or part of a financial debt that they owe to another country or to an international bank.

demography   The statistical study of human populations, including such information as numbers, distributions among race and ethnic group, and birth- and death rates.

development rights   Concessions that allow the conversion of land from natural area or agricultural use to residential, commercial, or industrial use.

discrimination   The practice of judging people on the basis of one or more classes to which they belong rather than on the basis of their individual characteristics.

discriminatory intent   A legal term that means that plaintiffs in an environmental justice (or other) case must be able to prove that race is a "motivating factor" in a decision made by an individual or corporation that brings harm to a person or group of people.

disproportionate   An adjective that refers to the fact that something is made available, distributed, or presented to people in a ratio that differs from the percentage of those people in the general population.

easement   A right granted by a landowner to another person or organization to use a piece of her or his land.

economic blackmail   A practice in which a corporation offers financial benefits to individuals or a community in exchange for putting up with environmental or other hazards.

effluent   Any material, usually a liquid, discharged from a point source into the surrounding environment.

emission standard   The maximum amount of any pollutant that is permitted by law or regulation to be released into the environment.

environmental discrimination   The disproportionate exposure to adverse environmental conditions as a result of racial, ethnic, economic, or other characteristics of a community.

environmental equity   A condition in which the burdens and benefits resulting from technological development are shared equally by all groups within society.

environmental high-impact area   An area that is subject to a higher-than-normal concentration of hazardous conditions, such as air and water pollution and/or hazardous waste sites.

environmental impact statement   A document that describes the effects that a proposed action is likely to have on the environment.

environmental justice   The attempt to achieve environmental equity for all groups within society.

environmental racism A term coined in 1982 by Benjamin F. Chavis meaning "racial discrimination in environmental policy making and the unequal enforcement of environmental laws and regulations" (U.S. Congress 1993).

environmental screening Tests performed to determine the exposure that an individual or a community has had to a hazardous substance.

environmentally disadvantaged community An area in which there exists at least one hazardous waste facility and that contains a higher-than-average percentage of low-income or minority residents.

environmentally sound management A term introduced in the Basel Convention on the Control of Transboundary Movements of Hazardous Wastes and Their Disposal to describe an environmentally correct way for handling hazardous waste.

external cost *See* **negative externality**.

fair share legislation A type of legislation that requires that environmental hazards be shared equally within a community and across communities.

grandfather right An exemption from a law or regulation because of the fact that a condition existed before the law or regulation was adopted.

grassroots A term describing any organization or movement that is derived from the most fundamental level of society; being organized from the bottom up, rather than from the top down.

growth management All processes involved in controlling the rate and circumstances under which new developments take place in a community.

hazardous waste Any solid, liquid, or gas released into the environment by an industrial process or from a municipal outlet that may cause damage to the health of a living organism.

host fee A payment made to individual homeowners or to the general fund of a community in return for its accepting the siting of a polluting or hazardous facility in its area.

hot spot An area that contains a dangerously high level of some hazardous material.

impact analysis A study conducted to find out how development will affect a particular area.

indigenous people People who are native to a particular area, region, or country.

institutional racism A form of racism that is maintained by a variety of legal and customary practices.

internal costs Costs of production that are paid by the producer or the consumer.

job blackmail   *See* **economic blackmail**.

laissez-faire   From the French, "allow to act," the principle of allowing people to act as they please.

litigation   Legal action.

LULUs (locally undesirable land uses)   Facilities such as polluting factories, hazardous waste dumps, and strip mines that make significantly negative environmental contributions to the region in which they are located.

mainstream environmentalism   A form of environmentalism that has grown up over the past century or more in which the emphasis has been on the conservation or preservation of natural resources or the solution of pollution problems.

move to the nuisance   A phrase that describes the tendency of people from lower economic groups to move into less-desirable living areas (such as those in which pollution is a serious problem) because housing prices are lower in those areas.

negative externality   A term used by economists to describe an unpleasant condition (such as the presence of air pollution) that is beyond the control of those people who are exposed to it. The condition is also known as an external cost.

negotiated compensation   A proposed method for dealing with the inequities of environmental racism by charging those entities that create environmental problems (such as the owners of a waste incineration plant) while compensating those who have to live in the area where the environmental degradation has taken place.

NIMBY (not in my backyard)   A phrase used to describe the opposition of individuals who oppose the siting of any environmentally harmful facility anywhere near the property they own or reside on.

NOPE (no place on Earth)   A somewhat pejorative term sometimes used to describe the position of "radical environmentalists" who oppose the construction of any facility that would produce pollution or other effects damaging to the environment anywhere on Earth.

nuisance law   Any regulation that prohibits a person or business from interfering with others' enjoyment and use of property.

offsetting benefits   Certain advantages given to a community in exchange for its accepting an environmentally hazardous facility in its area.

PIBBY (put in the blacks' backyard)   A policy, usually unspoken, by which facilities with adverse environmental effects are sited in neighborhoods occupied primarily by African Americans or other nonwhite groups.

point source Any readily identifiable location, such as a smokestack or sewage outlet pipe, from which pollution is released.

prior informed consent A term that means that a pesticide that has been banned, withdrawn, or severely restricted in one country cannot be exported to another country until and unless the second country has been informed of such action and has agreed to accept the pesticide under these conditions.

proof of intent A legal term that means that a complainant in a case must be able to show that a person or a company planned or knew that some damage would result from his, her, or its actions.

quality of life The extent to which a person is satisfied with his or her life.

rebuttable presumption An assumption about a situation that anyone is allowed to challenge.

Reserved Rights Doctrine A legal principle that states that rights mentioned in treaties between the U.S. government and Indian tribes were not granted by the government but were rights that belonged to Native Americans in the first place and that were only confirmed by the treaties.

right to inspect A policy under which industries extend to workers, neighbors, and others the right to enter and inspect their facilities to become more familiar with any potential environmental threat they may pose.

right to know A policy under which industries are required to provide information to workers, neighbors, and others exposed to hazardous environmental conditions about the nature of those conditions.

sacrifice zone An area in which there is an unusually high concentration of industries releasing pollutants to the surrounding region.

suitability analysis A study to find out if a piece of land should be used for some given purpose, such as urban development.

Superfund site Land that has been contaminated by hazardous waste and has been chosen by the Environmental Protection Agency (EPA) as a site for cleanup because of the risks it poses to human health.

sustainable development Development that takes place with minimal harmful effects on the physical and biological environment.

threshold effect A consequence that occurs only when a certain level of exposure has occurred.

Index

Note: t. indicates table.

Acid Rain Foundation, 20
Afghanistan, 95
Africa
  and Basel Convention, 95, 96
  and dumping of wastes from
    developed countries, 91
  and environmental inequities
    for Maasai, 85–86
African American Environmentalist
  Association (AAEA), 217
African Americans and
    environmental inequities
  "Cancer Alley" (Louisiana),
    9–11, 55–56
  consumption of toxic fish
    (Detroit River, Michigan),
    13–14
  diesel bus pollution (West
    Harlem, New York), 16–17
  and environmental activism
    and attitudes, 44–45
  PCB pollution (Sweet Valley
    and Cobb Town, Alabama),
    15–16
  and research of United Church
    of Christ, Commission on
    Racial Justice, 23–25, 38–39
  Warren County, North
    Carolina, 1–3, 23
  West Dallas, Texas, 6–7

Alabama, and environmental
    justice regulations, 73
Alaska, and environmental
    justice regulations, 63
Alaska Community Action on
    Toxics (ACAT), 235
Alexander v. Sandoval, 70–71
Alliance for Nuclear
    Accountability (ANA), 217
Alston, Dana Ann, 137–139
Alter, Harvey, 93
Alternatives for Community and
    Environment (ACE), 236
American Bar Association,
    Individual Rights and
    Responsibilities section,
    62–65
American Cyanamid, 88
Americans for Indian
    Opportunity (AIO),
    217–218
Anderton, Douglas L., 39–40
Arizona, and environmental
    justice regulations, 63
Asian Pacific Environmental
    Network (APEN), 218
Association of Southeast Asian
    Nations (ASEAN), 96
Atlanta Women's Action for New
    Directions, 17–18

About the Author

David E. Newton holds an associate's degree in science from Grand Rapids (Michigan) Junior College, a BA in chemistry (with high distinction) and an MA in education from the University of Michigan, and an EdD in science education from Harvard University. He is the author of more than 400 textbooks, encyclopedias, resource books, research manuals, laboratory manuals, trade books, and other educational materials. He taught mathematics, chemistry, and physical science in Grand Rapids, Michigan, for 13 years; was professor of chemistry and physics at Salem State College in Massachusetts for 15 years; and was adjunct professor in the College of Professional Studies at the University of San Francisco for 10 years. Previous books for ABC-CLIO include *Global Warming: A Reference Handbook* (1993), *Gay and Lesbian Rights: A Reference Handbook* (1994), *The Ozone Dilemma: A Reference Handbook* (1995), *Violence and the Mass Media: A Reference Handbook* (1996), *Encyclopedia of Cryptology* (1997), and *Social Issues in Science and Technology: An Encyclopedia* (1999).